16th International Conference on Organized Molecular Films (ICOMF16-LB16)

MATEC Web of Conferences Volume 98 (2017)

Helsinki, Finland
25 - 29 July 2016

Editor:

Tapani Viitala

ISBN: 978-1-5108-3608-2

Printed from e-media with permission by:

Curran Associates, Inc.
57 Morehouse Lane
Red Hook, NY 12571

Some format issues inherent in the e-media version may also appear in this print version.

Printed by Curran Associates, Inc. (2017)

For additional information, please contact EDP Sciences – Web of Conferences at the address below.

EDP Sciences – Web of Conferences
17, Avenue du Hoggar
Parc d'Activité de Courtabœuf
BP 112
F-91944 Les Ulis Cedex A
France

Phone: +33 (0) 1 69 18 75 75
Fax: +33 (0) 1 69 28 84 91

contact-edps@webofconferences.org

Additional copies of this publication are available from:

Curran Associates, Inc.
57 Morehouse Lane
Red Hook, NY 12571 USA
Phone: 845-758-0400
Fax: 845-758-2633
Email: curran@proceedings.com
Web: www.proceedings.com

TABLE OF CONTENTS

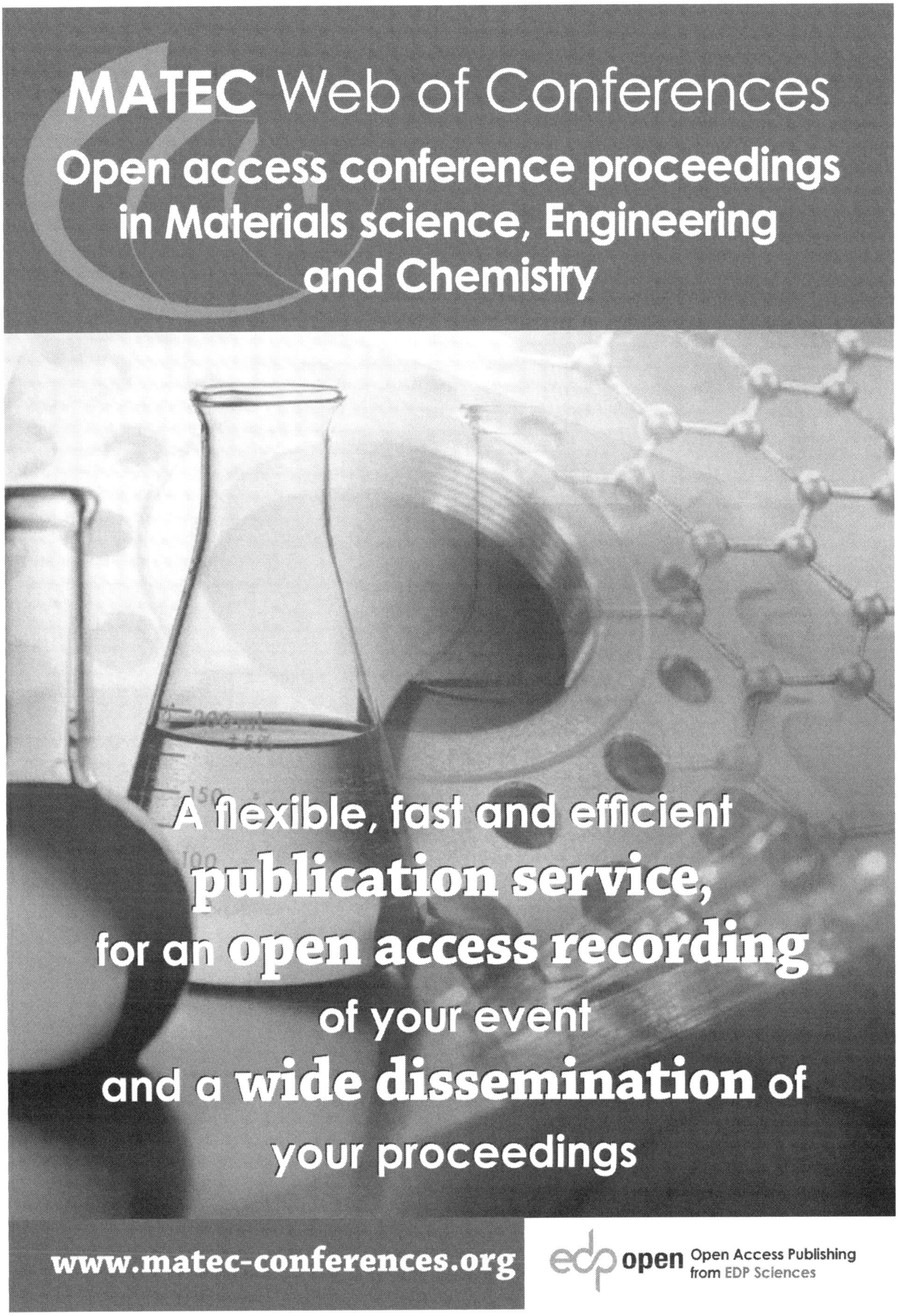
MATEC Web of Conferences
Open access conference proceedings in Materials science, Engineering and Chemistry
A flexible, fast and efficient publication service, for an open access recording of your event and a wide dissemination of your proceedings
www.matec-conferences.org
edp open Open Access Publishing from EDP Sciences

16th International Conference on Organized Molecular Films

25th – 29th July 2016, Helsinki, Finland

The ICOMF16 was the sixteenth international conference devoted to organized molecular films and nanostructures in general. The ICOMF16 was held in Helsinki, Finland during July 2016 and continued the tradition of highly successful biennial conferences hosted by many different countries on different continents. The 16th ICOMF-conference provided an excellent opportunity for chemists, physicists, biologists, material scientists, polymer scientists, chemical engineers and researchers in related fields to gather and discuss their latest results in the field of molecular thin films, nano-objects, nanostructures and bio-inspired systems, and to develop and extend their international collaborations.

Preface

The 16th International Conference on Organized Molecular Films was a continuation of highly successful biennial conferences organized in various countries and continents. The conference was for the first time organized in Nordic countries and the local organizing committee was delighted that Helsinki and Finland was given the opportunity to host this event. Especially, as the conference was organized on the advent of reaching a historical milestone, which has been the foundation for this conference series. Namely, in the year 2017 it will be 100 years since Irving Langmuir published his pioneering work on studying the properties of insoluble monolayers on the air-water interface (JACS 39 (1917) pp. 1848). Much has happened since then, and the local organizing committee is pretty sure that Irving Langmuir would have been delighted to see the legacy he has left behind in the field of organized molecular films.

Opening Address

Triggered by Langmuir´s and many others groundbreaking works on monomolecular films the ICOMF-LB conference series was established in Durham, UK in 1982 as a conference focusing solely on Langmuir-Blodgett Films. However, as with all scientific fields, research on organized molecular films advances fast and with the help of new fundamental knowledge the scientists continuously develop and produce new applications and innovations within this field. Thus, organized molecular films have become one of the cornerstones in many research fields such as fundamental chemistry and physics, nanoscience- and –technology, and biomimicking systems, just to mention a few. The cross-disciplinary nature of studying organized molecular films requires scientists from various fields to collaborate with each other. The ICOMF-LB conference is an excellent forum for chemists, physicists, biologists, material scientists, polymer scientists, chemical engineers, and researchers in related fields to gather and discuss their latest results in the field of molecular thin films, nano-objects, nanostructures, and bio-inspired systems.

Committees

International Advisory Committee

- Carlos Constantino, Brazil
- Christine DeWolf, Canada
- Xi Zhang, China
- Agnès Girard-Egrot, France
- Michel Goldmann, France
- Andreas Terfort, Germany
- Mitsumasa Iwamoto, Japan
- Masatsugu Shimomura, Japan
- Hoon-Kyu Shin, Korea
- Piotr Cyganik, Poland
- Rigoberto Advincula, USA

Local Organizing Committee

- Tapani Viitala, conference chair, University of Helsinki, Finland
- Marjo Yliperttula, University of Helsinki, Finland
- Elina Vuorimaa-Laukkanen, Tampere Univ. of Technology, Finland
- Helge Lemmetyinen, Tampere University of Technology, Finland
- Lasse Murtomäki, Aalto University, Finland
- Jouko Peltonen, Åbo Akademi University, Finland
- Leena Viitala, Secretary
- Frank Grunfeld, JFG Science and Education Ltd, United Kingdom

Proceedings Editor

Dr. Tapani Viitala

University of Helsinki, Faculty of Pharmacy

Division of Pharmaceutical Biosciences

P.O. Box 56

00014 Helsinki

Finland

E-mail: tapani.viitala@helsinki.fi

Sponsors

- Accurion GmbH, Göttingen, Germany
- Bionavis Ltd, Tampere, Finland
- Kibron Inc., Helsinki, Finland
- KSV NIMA, Helsinki, Finland
- Langmuir, ACS Journals, USA
- Doctoral Programme in Materials Research and Nanosciences (MATRENA), Helsinki, Finland
- Åbo Akademi Stiftelse, Turku, Finland
- Federation of Finnish Learned Societies, Helsinki, Finland

Statement of Peer review

In submitting conference proceedings to MATEC Web of Conferences, I certify to the Publisher that I adhere to the **Policy on Publishing Integrity** of the journal in order to safeguard good scientific practice in publishing.

1. All articles have been subjected to peer review administered by the proceedings editors.
2. Reviews have been conducted by expert referees, who have been requested to provide unbiased and constructive comments aimed, whenever possible, at improving the work.
3. Proceedings editors have taken all reasonable steps to ensure the quality of the materials they publish and their decision to accept or reject a paper for publication has been based only on the merits of the work and the relevance to the journal.

Title, date and place of the conference

16th International Conference on Organized Molecular Films (ICOMF16-LB16)
25th - 29th July 2016, Helsinki, Finland

Proceedings editor(s):

Dr. Tapani Viitala
Chairman of the ICOMF16-LB16 conference
University of Helsinki, Faculty of Pharmacy
Division of Pharmaceutical Biosciences
P.O. Box 56
00014 Helsinki
Finland
E-mail: tapani.viitala@helsinki.fi
Phone. +358-50-4154529

Date and editor's signature 27.1.2017

Tapani Viitala

The Role of Modifying Molecular Chains in the Formation of Organized Molecular Films of Organo-modified Inorganic Particles

Manami Iizuka, Yusaku Shidara, and Atsuhiro Fujimori*

Graduate School of Sci. and Eng., Saitama University, 225, Shimo-okubo, Sakura-ku, Saitama, 338-8570, Japan

Abstract The role of organo-modifying molecular chains in the formation of molecular films of organo-modified nanodiamond is discussed herein based on interfacial chemical particle-integration of organo-modified nanodiamond having a particle size of 5 nm. The surface of nanodiamond is known to be covered with a nano-layer of adsorbed water. This water nano-layer was exploited for organo-modification of nanodiamond with long-chain fatty acids via adsorption, leading to nano-dispersion of nanodiamond in general organic solvents as a mimic of solvency. The organo-modified nanodiamond dispersed "solution" was used as a spreading solution for depositing a mono-"particle" layer on the water surface, and a Langmuir particle layer was integrated at the air/water interface. Multi-"particle" layers were then formed via the Langmuir-Blodgett technique, and were subjected to fine structural analysis. The effect of organo-modification enabled integration and multilayer formation of inorganic nano-particles due to enhancement of the van der Waals interactions between the chains. That is to say, the "encounter" between the organo-modifying chain and the inorganic particles led to solubilization of the inorganic particles and enhanced interactions between the particles, which can be regarded as imparting new function to the organic molecules.

1 Introduction

Two-dimensional arrangement of nanoparticles has played an important role in the development of modern technology.[1] Arranged nanoparticles have potential application in numerous technical areas ranging from organic devices to biomaterials.[2] Furthermore, organic/inorganic hybrid materials[3] have captured the attention of scientists as well as engineers owing to their remarkably high dimensional stability and gas-barrier performance, in addition to their superior mechanical properties relative to conventional composite materials.[4]

It has been envisioned that two-dimensional arrangement of nanodiamond particles into various materials should effectively enhance several material functionalities. However, it is difficult to obtain regular arrangements of nanodiamond particles because the van der Waals interaction in inorganic materials is relatively weak. To overcome this limitation, surface modification using organic compounds is one prospective means of increasing the affinity between particles. Previously, formation and structure of molecular films comprising organo-modified clay are investigated.[5] In that study, the organo-modified inorganic material formed an extremely condensed monolayer on the surface of water. Moreover, a highly ordered layer structure along the *c*-axis and two-dimensional packing in the *ab*-plane have commonly been constructed by the Langmuir–Blodgett (LB) technique.[6] In addition, the authors have successfully fabricated highly ordered particle assemblies by the synthesis of long-chain fatty acid-modified zirconium oxide (zirconia) by using novel organo-modification methods.[7]

In the present study, high-density and two-dimensional arranged nanodiamond organization is performed by an interfacial chemical method. Ultrathin organo-modified nano-diamond films are constructed by floating Langmuir monolayers. The technique proposed in this study is based on the nanodispersion of insoluble inorganic particles by solubilization in a general organic solvent. In general, the outermost layer of the nanodiamond with single-nanometer diameter is covered by a nano-layer of water for structural stabilization. The challenge of organo-modification of inorganic particles through the surface nano-phase of water is addressed herein. This nanodispersion of fine inorganic particles can be used as a "spreading solvent" for Langmuir monolayers. In other words, the formation of LB multilayers of nanodiamonds is made possible by using the proposed technique. In this study, LB multilayers of organo-modified nanodiamond films are characterized by out-of-plane and in-plane X-ray diffraction (XRD) and atomic force microscopy (AFM).

2 Experimental Section

2.1 Synthesis and characterization of organo-modified nanodiamond.

A dispersion was prepared by combining an aqueous solution of nanodiamond (detonated nanodiamond, New Metals and Chemicals Co., Ltd., particle size: 4.8±0.7 nm) with a methanolic solution of stearic acid. Toluene was poured into the nanodiamond dispersion with stirring. In this step, the organo-nanodiamonds migrated from the methanolic dispersion into the toluene phase. Water, methanol, and the remaining unreacted reagent were then removed by rotary evaporation under reduced pressure and decantation. This process changed the surface absorbed hydronium ions ($R\text{-}H_2O^+$) in the water nano-layer on the particle surface to long-chain alkyl carboxylates, as indicated by IR analyses. It is well known that the nanodiamond surface is coated with a water layer that stabilizes the structure. In this case, it is proposed that carboxylic ions (COO^-) are adsorbed by the hydronium ions on the nanodiamond surface.

2.2 Formation of monolayers on water surface and estimation of molecular arrangement in films.

* Corresponding author: fujimori@fms.saitama-u.ac.jp

It is possible to achieve uniform size dispersion of the generated organo-nanodiamond on the nanometer scale, where this process can essentially be considered as dissolving organo-nanodiamond in an organic solvent. Therefore, it was also possible to prepare a spreading solvent for formation of a monolayer on the water surface of the nanodiamonds. A Monolayer of organo-nanodiamond was formed by spreading a toluene solution containing organo-nanodiamond (ca. 1.0×10^{-4} M) on ultra-pure water (18.2 MΩ·cm). The concentration of organo-nanodiamond solutions was estimated by using the average weight of the modified-chain molecules calculated on the basis of the chemical characterization described above. After evaporating the toluene for 30 min, surface-pressure–area (π–A) isotherms were recorded at a compression speed of 0.08 mm·s^{-1}. The temperature of the air/water interface was maintained at a constant value of 15 °C by circulating thermostatically controlled water around the trough. Measurements of the monolayer properties and LB film transfer were carried out using a USI-3-22 Teflon-coated LB trough (USI Instruments). The monolayers were transferred onto mica (AFM samples) or glass substrates with a ferric stearate monolayer as a hydrophobic underlayer (out-of-plane XRD and in-plane XRD) at 15 °C, 0.2 mm·s^{-1}, and 25 mNm^{-1} using the LB method.

2.3 Surface morphology and particle arrangement in organized films.

The surface morphology of the transferred monomolecular films was observed using a scanning probe microscope (Dynamic Force Mode, Scan speed : 2.0 Hz, Seiko Instruments, SPA300 with a SPI-3800 probe station) and microfabricated rectangular Si cantilevers with integrated rectangular tips; a spring constant of 1.4 N·m^{-1} was applied in this process. The large spacing between the layer structures of the films transferred onto the glass substrates was measured using an out-of-plane X-ray diffractometer (Rigaku, Rint-Ultima III; Cu*K*α radiation, 40 kV, 40 mA) equipped with a graphite monochromator. The in-plane spacing of the two-dimensional lattice of the films was determined using an X-ray diffractometer with different geometrical arrangements[8, 9] (Bruker AXS, MXP-BX; Cu*K*α radiation, 40 kV, 40 mA, an instrument especially made to order) that was equipped with a parabolic graded multilayer mirror. X-rays were incident at an angle of 0.2° and the films were slow-scanned at a speed of 0.05°/80 sec.

3 Results and discussion

The interfacial monolayer behavior of organo-modified inorganic fine particles is extremely interesting in terms of elucidation of whether the inorganic monolayer is strongly influenced by any of the components. Figure 1(a), (b), (c), and (d) show the π–A isotherm of a single particle layer on the water surface of organo-nanodiamond and stearic acid as an organo-modifying agent. Figure 1(a) and (b) show a comparison of the monolayer behavior of organo-modified nanodiamond, stearic acid, and Cd stearate. The horizontal axes in Fig. 1(a) and (b) respectively indicate the "Area" and "Compression percentage", where the "Area" corresponds to "area/molecule" in the case of

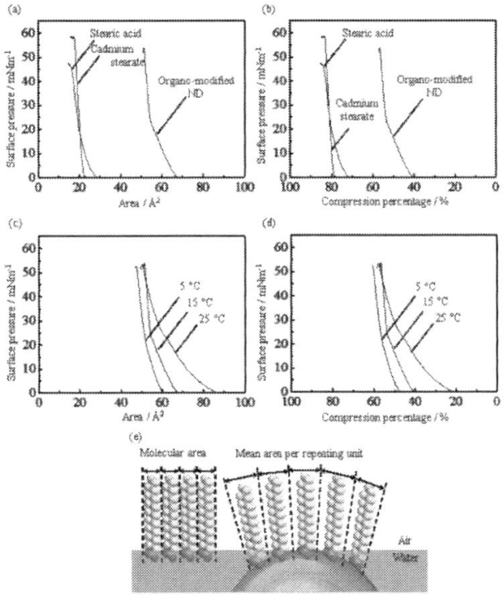

Fig. 1. (a) Surface pressure-area (π-A) and (b) Surface pressure-compression percentage isotherms of monolayers of several materials on the water surface (15 °C). (c) π-A, and (d) π-compression percentage isotherms of single-particle layers of organo-modified nanodiamond on the water surface (5, 15, and 25 °C). (e) Illustration of surface modified agent at air/water interface.

stearic acid and stearic acid cadmium, and indicates the "mean area per hydrophobic repeating unit" in the case of nanodiamond. In this case, the concept of molecular weight cannot be applied because the total number of atoms of nanodiamond particle units cannot be exactly determined. Therefore, considering the surface conformation as shown in Fig. 1(e), the repeating units of only the stearate group of the organo-nanodiamond surface are shown in Fig. 1(a) and (c). However, it is expected that the most accurate comparison would be derived by using the "compression percentage", which indicates the mean of the compression ratio versus the total surface area of the LB trough, as the horizontal axis. Notably, the π-A curve of organo-modified nanodiamond is clearly different from that of stearic acid and Cd stearate. From the results of the π-A isotherms of the organo-modified ND monolayer, both expanded and condensed phases were confirmed in the low and high surface pressure regions, respectively.

In addition, two-dimensional phase transition from the expanded phase to the condensed phase is indicated by the changes in the inclination of π-A curve of organo-nanodiamond monolayer. The appearance of this phase transition is an effect derived from the organic component, while the appearance of a clear crystalline phase in the high-pressure region is believed to be an effect of the inorganic component. The temperature dependence of the π-A curve was evaluated to confirm the occurrence of the effect derived from the organic component (Fig. 1(c) and (d)). Specific condensation and expansion behavior typical of the π-A isotherms of organic compounds was confirmed with the change of the subphase temperature.

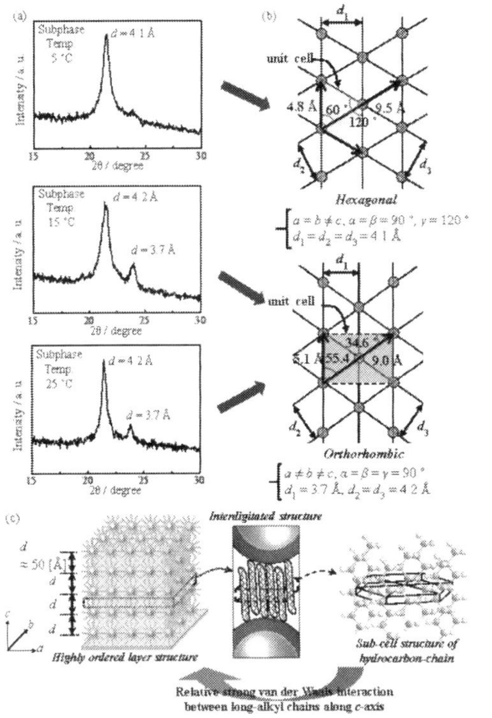

Fig. 2. (a) Mesoscopic AFM images of Z-type single-particle layer of organo-modified nanodiamond transferred by LB method (5, 15, and 25 °C). (b) Out-of-plane XRD profiles of multi-particle layers of organo-modified nanodiamond transferred by LB method (5, 15, and 25 °C). (c) Schematic illustration of multi-particle layered organization of organo-modified nanodiamond. (d) Photograph of the occurrence behavior of structural colors of step-wise multilayers of organo-modified nanodiamond.

This single-particle layer on the water surface was transferred to a mica substrate at various temperatures by using the upstroke LB method. Figure 2(a) shows AFM images of these Z-type single-particle layers. Formation of an organized molecular film of organo-modified nanodiamond was also confirmed by Laser Raman spectroscopy. A diamond-specific 1330 cm^{-1} Raman shift band[10] was confirmed in the Raman spectrum of the organized film, in addition to the C-C band at 1098 cm^{-1}. In a nutshell, it was not possible to obtain a quite homogeneous single particle film at this stage. In particular, because the height is about three times the particle diameter in the case of the layer transferred at 5 °C, it seems that there are sites where the particles aggregate and pile up along the height direction. In the case of the single-particle layers transferred at 15 and 25 °C, although the height direction is homogeneous and consistent with single particle size, the tendency towards particle integration in the in-plane direction is not necessarily high. This problem is eliminated by the "repeating compression-expansion method" discussed in an ensuing section.

On the other hand, the formation of a quite highly ordered layer arrangement along the height direction was inferable

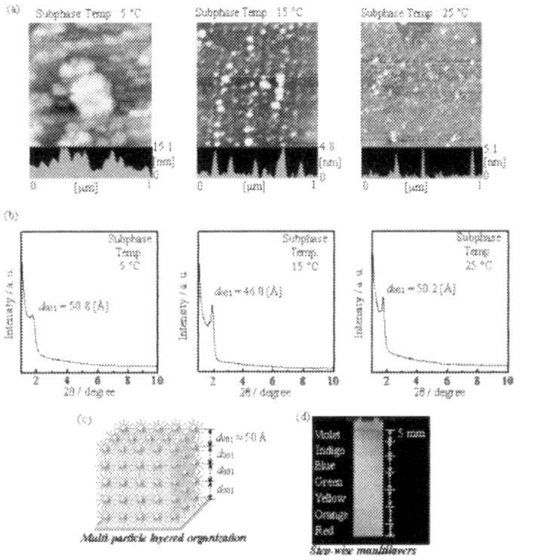

Fig. 3. (a) In-plane XRD profiles of multi-particle layers of organo-modified nanodiamond transferred by LB method (5, 15, and 25 °C). (b) Schematic models of two-dimensional lattices formed by long hydrocarbons on organo-modified nanodiamond surface of their LB films transferred at several subphase temperature. (c) Schematic illustration of formation of sub-cell structure in high-order layered organization of organo-modified nanodiamond based on the contribution of strong van der Waals interaction between alkyl chains.

from the data in Fig. 2(b). The results of out-of-plane X-ray diffraction of the LB multilayers indicates that an organized film was transferred at all subphase temperatures, and the films showed a clear long-period value at around 5 nm. This value corresponds to the particle lamination period as shown in Fig. 2(c). The diffraction peaks in transferred at 15 °C and transferred at 25 °C were comparatively sharper than that in transferred at 5 °C; therefore, more advanced layered structures were formed in the former films. As further evidence, the multilayers fabricated stepwise at 15 °C exhibited a clear structural related color up to wavelengths in the visible light region in response to stacking (Fig. 2 (d)). In this case, gradation of structural color has been indicated on the stepwise-multilayers. The violet, indigo, blue, green, yellow, orange, and red colors correspond to 38, 44, 49, 54, 60, 65, and 70 layers of organo-nanodiamond, respectively. Figure 3 shows the in-plane X-ray diffraction profiles of multi-particle layers of organo-nanodiamond that were transferred at several temperatures. Unfortunately, the diffraction peak in the low angle region corresponding to the long spacing in the out-of plane XRD pattern was not observed. Based on the AFM results, this observation may be reasonable. The arrangement of the particles in the two-dimensional plane would not be highly regular at this stage. However, the modifying long-chain alkyl group was found to have a high degree of order in the particle layers. A single peak corresponding to the isotropic hexagonal (100) plane was apparent at 4.1 Å in the case of the multi-particle layers fabricated at the subphase temperature of 5 °C. Additionally, the appearance of double peaks in the case

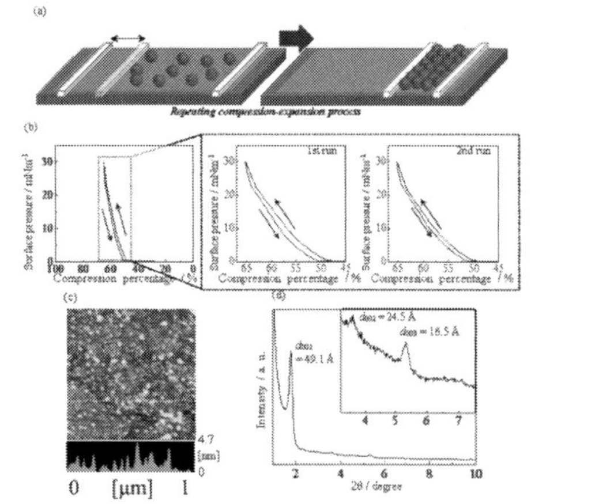

Fig. 4. (a) Schematic illustration of "repeating compression-expansion method". (b) Hysteresis curves of π-A isotherms of single-particle layers of organo-modified nanodiamond on the water surface (15 °C). (c) AFM image of single-particle layer of organo-modified nanodiamond fabricated by "repeating compression-expansion method" at 6 times. (d) Out-of-plane XRD profiles of multi-particle layers of organo-modified nanodiamond performed "repeating compression-expansion method" at 6 times.

of the multi-particle layers fabricated at the subphase temperatures of 15 and 25 °C indicates that the modified alkyl chains form a slightly anisotropic orthorhombic system. In view of the ratio of surface modification of the particles by stearic acid, the possibility that the long alkyl chain is densely packed on one particle surface is extremely low. Accordingly, packing of the alkyl chains is expected to generate a two-dimensional crystal array of the alkyl chains between individual particles. In other words, the two-dimensional integration and formation of the layer structure of organo-nanodiamond can be determined, indicating that crystal formation arises from interaction between modified alkyl chains. This inferred model is represented in Fig. 3(c). Sub-cell[11] formation by the hydrocarbons between the particles is considered to be the origin of the nano-particle integrated organization in this system.

As the next step, the formation of the high-density and low-defect particle layer was addressed. Figure 4(a) shows a schematic illustration of the "repeating compression-expansion process" for generation of the single particle layer on the water surface. In a system where the size of the components is large and the interaction between constituent units is relatively weak, a low-defect layer can be effectively formed from this process. The fact that this technique works effectively reflects the monolayer properties of the inorganic compound. In the case of an organo-modified montmorillonite single particle layer, this technique was effectively applied to fabricate low-defect films. Figure 4(b) illustrates the reversibility and hysteresis of the π-A isotherm at 15 °C. At each step of the repeating process, the difference in the hysteresis during compression-expansion becomes less. When this process is performed 6 times, the 5th and 6th curves of the hysteresis loop almost coincide. Therefore, considering that the highest density films had successfully been attained, the resulting monolayer and LB multilayers transferred to the solid substrate were subjected to AFM observation and out-of plane X-ray diffraction. The results confirm a very low-defect surface morphology (Fig. 4(c)). In addition, the d_{001} peak in the out-of plane X-ray diffraction pattern became very sharp (Fig. 4(d)). Comparing the magnitude of the D_{001} for the crystallites of the fabricated films based on the full width at half maximum before and after performing the sequence of the "repeating compression-expansion process" for six cycles, the diameter of the crystallites clearly increased from 364 Å to 461 Å (the Scherrer constant $K = 0.94$ was used).

Conclusions

The role of organic chain molecules used for surface modification of inorganic particles was discussed *via* construction of two-dimensional integration and layered lamination of organo-modified nanodiamond. Multi-particle layered films of organo-nanodiamond were constructed by using the LB technique. Mesoscopic-scale AFM images of this single-particle layer of organo-nanodiamond fabricated by the "repeating compression-expansion process" demonstrated formation of a high-density surface with a uniform height of 5 nm. These regular structures are based on the strong van der Waals interactions of the organo-modifying chains.

References

1. Y. Yin, Y. Lu, B. Gates, Y. Xia, J. Am. Chem. Soc., **123**, 8718 (2001)
2. A. Fujimori, N. Sato, S. Chiba, Y. Abe, Y. Shibasaki, J. Phys. Chem. B, **114**, 1822 (2010)
3. K.M. Kim, D.K. Keum, Y. Chujo, Macromolecules, **36**, 867 (2003)
4. A. Fujimori, N. Ninomiya, T. Masuko, Polym. Eng. Sci., 48, 1103 (2008)
5. A. Fujimori, J. Kusaka, R. Nomura, Polym. Eng. Sci., **51**, 1099 (2011)
6. T. Yamamoto, Y. Umemura, O. Sato, Y. Einaga, J. Am. Chem. Soc., **127**, 16065 (2005)
7. A. Fujimori, N. Honda, H. Iwashita, Y. Kaneko, S. Arai, M. Sumita, S. Akasaka, Colloids Surf. A, **446**, 109 (2014)
8. A. Fujimori, Y. Sugita, H. Nakahara, E. Ito, M. Hara, N. Matsuie, K. Kanai, Y. Ouchi, K. Seki, Chem. Phys. Lett., **387**, 345 (2004)
9. A. Fujimori, T. Araki, H. Nakahara, E. Ito, M. Hara, H. Ishii, Y. Ouchi, K. Seki, Chem. Phys. Lett., **349**, 6 (2001)
10. O.A. Shenderova, I.I. Vlasov, S. Turner, G.V. Tendeloo, S.B. Orlinskii, A.A. Shiryaev, A.A. Khomich, S.N. Sulyanov, F. Jelezko, J. Wrachtrup, J. Phys. Chem. C, **115**, 14014 (2011)
11. V. Vand, I.P. Bell, Acta. Cryst., **4**, 465 (1951)

Structural Characterization of Self-Organized Mono- and Multilayers of Partly Fluorinated Polydialkoxyphosphazenes at the Air/Water Interface

A.I. Buzin[1,*], *G.* Brezesinski[2], *V.S.* Papkov[3], and *S.N.* Chvalun[4,1]

[1]Enikolopov Institute of Synthetic Polymer Materials Russian Academy of Sciences, ul. Profsoyuznaya 70, 117393 Moscow, Russia
[2]Max Planck Institute of Colloids and Interfaces, Am Mühlenberg 1, 14476 Potsdam, Germany
[3]Nesmeyanov Institute of Organo-Element Compounds Russian Academy of Sciences, Vavilova ul. 28, 117991 Moscow, Russia
[4]National Research Centre "Kurchatov Institute", Ak. Kurchatova pl. 1, 123182 Moscow, Russia

Abstract. We report studies of the collapse of monolayers of mesomorphic poly[*bis*-(2,2,3,3,4,4,5,5-octafluoropentoxy)phosphazene] and poly[*bis*-(2,2,3,3,4,4,5,5,6,6,7,7-dodecafluoroheptoxy)phosphazene]. The structure of the films has been examined at different stages of the collapse process by synchrotron grazing incidence X-ray diffraction. Surface pressure–surface area isotherms show that these films undergo a stepwise collapse with formation of multilayers. Grazing incidence X-ray diffraction reveals only one diffraction peak for the monolayer as well as for multilayers of the two polymers studied, which corresponds to an in-plane interchain spacing d = 15.2±0.1°Å and 17.6±0.5°Å, respectively. The in-plane correlation length ξ_{xy} reaches *ca.* 500 Å for **PFP-C5** (33 macromolecules) and *ca.* 400 Å for **PFP-C7** (23 macromolecules), implying a certain mesoscale order similar to a quasi-long-range one.

1 Introduction

Polyphosphazenes are an important class of organo-element polymers as they possess not only a number of properties useful for technological applications, but also exhibit some interesting structural features [1, 2]. Many of these polymers are capable of forming thermotropic mesophases, despite the fact that they possess a flexible main chain and do not contain any mesogenic fragments neither in the main nor in the side chains. These mesophases, named as *condis*-crystal or hexagonal columnar mesophase, are partly disordered crystalline modifications in thermodynanic equilibrium. They are formed by parallel packed macromolecules in extended macroconformation, main and side chains being conformationally disordered [3, 4]. Similar types of mesophases were observed in some linear and cyclolinear polyorganosiloxanes [3, 4].

The phase behavior in bulk of poly[*bis*-(2,2,3,3-tetrafluoropropoxy)phosphazene] (**PFP-C3**), poly[*bis*-(2,2,3,3,4,4,5,5-octafluoropentoxy)phosphazene] (**PFP-C5**) and poly[*bis*-(2,2,3,3,4,4,5,5,6,6,7,7-dodecafluoroheptoxy)phosphazene] (**PFP-C7**) was studied in Refs. 5 and 6, which was shown to be determined by an interplay between the packing of macromolecules on the whole and the registering of side groups. All three polymers form low-temperature crystalline phase and *condis*-mesophase. **PFP-C3** and **PFP-C5** also form second crystalline phase, which was interpreted to have a close to hexagonal packing of only side groups. It is an intermediate phase between the low-temperature 3D crystalline and mesomorphic phases. **PFP-C7** does not form a similar intermediate phase in bulk.

In this respect the above set of polyphosphazenes with partially fluorinated side alkoxy substituents of different lengths seem to be a suitable object for a more complete understanding the main features of various 2D and 3D ordered structures formation affected by the side groups interaction and arising due to consecutive discrete macromolecules layering which is realized at Langmuir monolayers collapse.

Up to date, only a few results were published concerning the behavior of Langmuir monolayers of polyphosphazenes particularly including partly fluorinated ones. Reversible formation of monolayers of poly[*bis*-(2,2,2-trifluoroethoxy)phosphazene] at the air/water interface, which can be successfully transferred onto solid substrates imparting a very high hydrophobicity to their surface was found in Ref. 7. The layering transitions of ultra-high MW **PFP-C3** induced by compression were found to result in the formation of stable multilayers directly at the air/water interface [8]. Their structure was characterized by synchrotron X-ray methods. The

* Corresponding author: al37919@gmail.com

transition into a 3D crystalline triclinic phase was observed after formation of the 4-layer thick film.

In this work we present results of the structural investigation of ultrathin layers of **PFP-C5** and **PFP-C7** performed *in situ* directly at the air/water interface using synchrotron X-ray diffraction, complementing published earlier corresponding data on **PFP-C3** and discuss the generality and differences in the structural changes in Langmuir monolayers of these polymers under increasing surface pressure.

2 Materials and Methods

2.1 Samples

Poly[*bis*-(2,2,3,3,4,4,5,5-octafluoropentoxy)phosphazene] ([η] = 5.45 dL g^{-1}, M_η = 8.1 10^6 Da), and poly[*bis*-(2,2,3,3,4,4,5,5,6,6,7,7-dodecafluoroheptoxy)phosphazene] ([η] = 3.46 dL g^{-1}, M_η = 6.8 10^6 Da) were synthesized in INEOS RAS in the group of Dr. D. R. Tur [6].

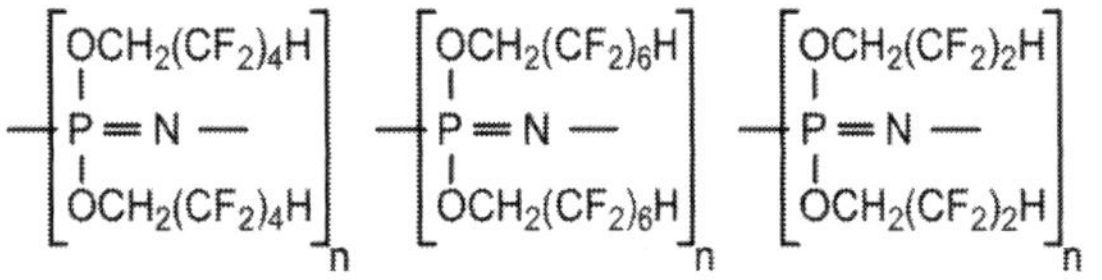

PFP-C5 **PFP-C7** **PFP-C3** [8]

2.2 Surface pressure–surface area isotherms

The surface pressure (π) *vs.* surface area (*A*) isotherms were recorded with a R&K (Potsdam, Germany) trough system. equipped with a Wilhelmy-type pressure sensing system. The velocity of compression was *ca.* 4 Å^2 (monomer unit)$^{-1}$ min^{-1}. The polymer was spread onto the subphase surface from ethylacetate solution with a typical concentration of 1 mg mL^{-1}. Water used as a liquid subphase was purified with a Millipore desktop system leading to a specific resistance of 18.2 MΩ cm. The subphase temperature was maintained at 20±0.5 °C.

2.3 Synchrotron grazing incidence X-ray diffraction

Synchrotron grazing incidence X-ray diffraction (GIXD) data collection was performed with a liquid surface diffractometer at the undulator beamline BW1 which was located in the synchrotron radiation research laboratory HASYLAB at DESY (Hamburg, Germany) [9, 10]. GIXD measurements were performed using a Langmuir trough with an area of 480 cm^2 placed in an air-tight aluminum container with Kapton windows. Surface pressure was measured with a Wilhelmy filter paper plate with accuracy ±0.2 mN m^{-1}. A polished glass block in the trough reduced the depth of the liquid subphase to 0.3 mm in the X-ray footprint region and thereby suppressed mechanically excited capillary waves. X-ray experiments were conducted under a water vapor saturated He atmosphere.

GIXD experiments were performed as follows. A monochromatic X-ray beam (λ = 1.304 Å) was selected with

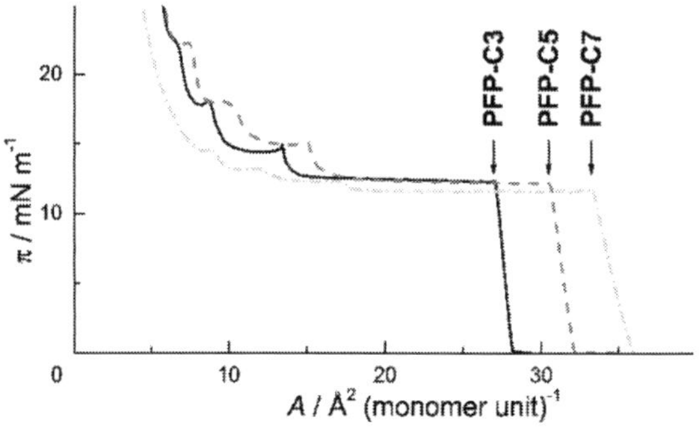

Fig. 1. Surface pressure–surface area isotherms of **PFP-C5**, **PFP-C7** and of **PFP-C3** [8] at 20 °C.

a Be(002) monochromator crystal. The incident grazing angle α_i of the X-ray beam was ~0.85 α_c, where α_c = 0.129° is the critical angle for total external reflection from an air/water interface. This limits the penetration depth of the beam to ~8 nm [9]. Thus, X-ray scattering from the water subphase contributing to the background intensity is efficiently reduced allowing accurate measurements of the X-ray photons diffracted from the monolayer. The dimension of the X-ray footprint on the sample was 2 mm×50 mm. A linear position sensitive detector (PSD) (OED-100-M, Braun, Garching, Germany) with a vertical acceptance $0 < \alpha_f < 15°$ scanned over $2\theta_{xy}$ in the range $0 < 2\theta_{xy} < 40°$ was used for recording the diffracted intensity. The measured GIXD data can be visualized as contour plots with the intensity shown as a function of both the horizontal (q_{xy}) and the vertical (q_z) scattering vector components or projected onto the q_{xy} or q_z axes as Bragg peak or Bragg rod data, respectively. The horizontal resolution of 0.008 Å^{-1} was determined by a Soller collimator mounted in front of the PSD. The lattice spacing d_{hk} can be obtained from the in-plane diffraction data as $d_{hk} = 2\pi / q_{xy}^{hk}$, where the Bragg peaks have been indexed by the Miller indices *h*,*k* to yield the unit cell parameters for the in-plane lattice. The full-width at half-maximum (*fwhm*) of the Bragg peaks Δq_{xy} yields the in-plane correlation length $\xi_{xy} \approx 0.88 \times 2\pi / \Delta q_{xy}$. The *fwhm* of the Bragg rods Δq_z gives a first estimate of the film thickness: $L \approx 0.88 \times 2\pi / \Delta q_z$ [11].

3 Results and discussion

3.1 General characterization of the π–*A* isotherm

The polymers investigated are able to form monolayer films on the water surface. Surface pressure–surface area isotherms at 20 °C of **PFP-C5**, **PFP-C7**, as well as of **PFP-C3** taken from Ref. 8 for comparison are shown in figure 1. The parameters of the π–*A* isotherms are summarized in table 1. At large areas per monomer unit the surface pressure remains immeasurably low. In this region the condensed monolayer domains coexist with the bare water surface regions. On compression beyond 32.2 Å^2 (**PFP-C5**), and 36.0 Å^2 (**PFP-C7**) per monomer unit the surface pressure increases linearly up to the beginning of the first plateau. In this region occurs elastic deformation of the dense polymer monolayer of polymer.

Table 1. Parameters of the π–A isotherms of **PFP-C5** and **PFP-C7** (n — step number).

n	A_n (Å^2)	π (mN m^{-1})	A_n / A_1
	PFP-C5		
1	30.5	12.2	
2	15.2	14.9	0.5
3	10.4	17.9	0.34
4	7.6	22.1	0.25
	PFP-C7		
1	33.4	11.6	
2	17.0	12.3	0.51
3	12.0	13.2	0.36
4	9.1	14.5	0.27

Further compression of the surface layer of the both polymers studied is accompanied by four distinct steps in the surface pressure. The ratios of the surface areas of the steps A_n to the surface area of the dense monolayer A_1 for **PFP-C5** are close to 1/2, 1/3 and 1/4. Such area ratios allow to suggest the formation of the series of discrete multilayers (bilayer, trilayer, tetralayer) similar to that observed earlier in **PFP-C3** [8] and mesomorphic cyclolinear polyorganosiloxanes [12–14]. In the case of **PFP-C7** the area ratios are slightly overvalued, which may reflect distortion of the shape of the macromolecule in the multilayers (see GIXD data). One can see that **PFP-C5** demonstrate about the same surface pressure values as **PFP-C3** and only possesses larger molecular area due to longer side groups. At the same time, **PFP-C7** shows lower surface pressure of the first step as well as average surface pressure of the following steps *ca.* 1 mN m^{-1} *vs.* 3–4 mN m^{-1}, which is typical for **PFP-C3** and **PFP-C5**. This may indicate lower stability of the **PFP-C7** layer in comparison with **PFP-C3** and **PFP-C5**.

It is important to note that the surface pressure–surface area isotherms are not equilibrium ones. On decompression of the multilayer the surface pressure drops to a value slightly lower than the pressure of the first plateau and stays constant. The surface pressure decreases after stopping the compression and increases after stopping the expansion. An equillibrium surface pressure value exists for every plateau region. The reversibility and stability of the π–A isotherm of **PFP-C3** was discussed in detail in Ref. 8. **PFP-C5** and **PFP-C7** show similar behavior.

3.2 Structure of mono- and multilayers

Partly fluorinated polydialkoxyphosphazenes demonstrate the thermotropic polymorphism, *i.e.*, the 3D low-temperature and intermediate crystalline phases and one mesophase form in the bulk state [5, 6]. They transform reversibly one into another on heating and cooling. In **PFP-C3** the crystal-crystal transition occurs in a temperature range from 5 to 30 °C depending on the thermal history of the sample, whereas the crystal-mesophase transition occurs above 80 °C. In **PFP-C5** these two transitions are observed at lower temperatures (8 °C and 45 °C, respecttively). **PFP-C7** on heating transforms from low-temperature crystalline phase directly into mesophase at 0 °C. The low-temperature crystalline Phase I is characterized by the common crystalline lattice for the main chain and side chains packed into a monoclinic unit cell. The intermediate crystalline Phase II is controlled mainly by close to hexagonal packing of the side chains. The transition into the mesomorphic phase is accompanied by a distortion of the side chains packing. Only regular packing of the main chains of macromolecules in the plane perpendicular to their axes remains in the mesomorphic state.

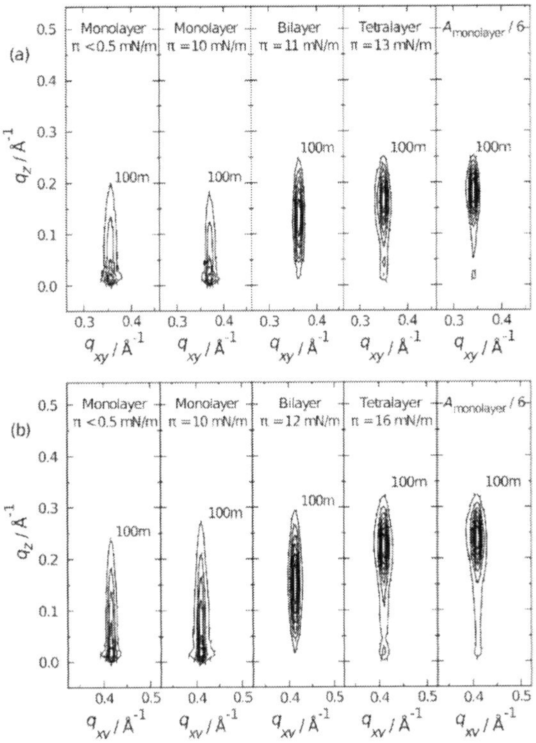

Fig. 2. Contour plots of the scattered intensity as a function of the in-plane q_{xy} and out-of-plane q_z scattering vector components of **PFP-C5** (a) and **PFP-C7** (b) mono-, bi-, tetralayers at various surface pressures (indicated) as well as at 1/6 of the dense monolayer area.

Selected GIXD data for thin films of **PFP-C5** and **PFP-C7** are shown in figure 2 as 2D contour plots. The scattering from the monolayer shows only a single Bragg peak related to the in-plane interchain distance. The absence of any further diffraction peaks, and therefore, the absence of any long-range correlation along the macromolecules both in monolayers, and in the collapsed films, can most likely be explained as a consequence of a conformational disorder and/or non-registering of one polymer chain along the next one, that is typical for the *condis*-mesophase in the bulk. In-plane repeating distance d and correlation length ξ as well as layer thickness L obtained from the GIXD experiments for the monolayers and the collapsed mesomorphic films of **PFP-C5**, **PFP-C7** and for comparison of **PFP-C3** [8] are shown in figure 3.

Monolayers exhibit a single diffraction peak already at a very low surface pressure (less than 0.5 mN m^{-1}) just before the beginning of the first increase in π. The position of the Bragg peak gives an in-plane repeating distance d in the monolayer (d = 15.2 Å for **PFP-C5** and 17.6 Å for **PFP-C7**). These values are slightly lower than the interchain distances (d = 15.7 Å and 18.1 Å, respecti-

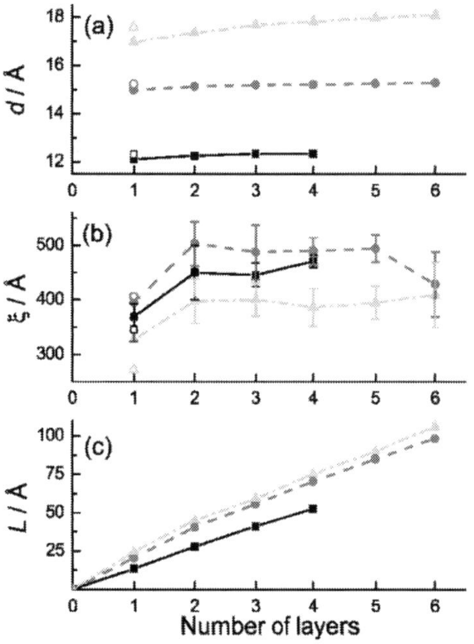

Fig. 3. In-plane repeating distance *d* and correlation length ξ as well as layer thickness *L* obtained from the GIXD experiments for the mesomorphic surface layers of **PFP-C5** (circles), **PFP-C7** (triangles) and **PFP-C3** (squares) [8]. Open symbols indicate points measured at $\pi < 0.5$ mN m^{-1}. Errors in *d* and *L* are smaller than symbol size.

vely) measured in the bulk *condis*-mesophase with hexagonal columnar packing of the main chains [6]. Compression of the monolayer is accompanied by a slight decrease in the repeating distance (at 10 mN m^{-1} $d = 15.0$ Å and 17.0 Å, respectively). With increasing number of layer, the in-plane repeating distance of **PFP-C5** stays constant at *ca.* 15.2 Å, while the in-plane repeating distance of **PFP-C7** grows reaching 18.1 Å for 1/6 of the dense monolayer area. Analysis of the observed *fwhm* of the Bragg peak Δq_{xy} yields the in-plane correlation length ξ_{xy}, *i.e.*, the extent of lateral positional correlations within the monolayer and collapsed films, perpendicular to the axes of the macromolecules. Starting from bilayer the correlation length reaches *ca.* 500 Å for **PFP-C5** (33 macromolecules) and *ca.* 400 Å for **PFP-C7** (23 macromolecules). The analysis of *fwhm* of the Bragg rod allows estimating the thickness *L* of the mono- and multilayers. The monolayer thickness is 20.7 Å for **PFP-C5** and 24.6 Å for **PFP-C7**, thus indicating essential distortion of the cylindrical shape of the macromolecules in the monolayer. With increasing layer numbers, the layer thickness grows non-linearly. At 1/6 of the dense monolayer area $L = 98.3$ Å for **PFP-C5** and 106.1 Å for **PFP-C7**, that means approaching the cylindrical shape of the macromolecules, which is typical for bulk hexagonal columnar mesophase.

4 Conclusion

We investigated structure of mono- and multilayers of mesomorphic partly fluorinated polydialkoxyphosphazenes at the air/water interface by GIXD. The scattering from both the monolayer and the multilayers shows only a single Bragg peak, which indicates regular packing in the direction perpendicular to the axes of the macromolecules. The macromolecules are mainly in extended chain conformation and their main chains are aligned parallel to each other. The periodicity along the macromolecules is largely disrupted due to the conformational disorder and non-registering in the main chains, that is typical for the *condis*-mesophase in the bulk state. The extent of lateral correlations between parallel macromolecules ranges from 33 (for **PFP-C5**) to 23 (for **PFP-C7**) interchain distances, demonstrating a mesoscale-range order for the collapsed films, which resembles the 2D order in the columnar mesophase. The monolayer collapse of polyphosphazenes studied looks similar to that observed earlier for some cyclolinear polyorganosiloxanes [12] suggesting that the ability of organo-element polymer to form columnar mesophase can be a prerequisite of the phenomenon of multilayer formation.

Acknowledgements

This work was supported by Russian Science Foundation (project 14-13-01402). We are grateful to HASYLAB at DESY, Hamburg, Germany for beam time.

References

1. H.R. Allcock, *Chemistry and applications of polyphosphazenes* (Wiley-Interscience, 2003)
2. A.K. Andrianov, *Polyphosphazenes for biomedical applications* (John Wiley & Sons, 2009)
3. Y.K. Godovsky, V.S. Papkov, Adv. Polym. Sci. **88**, 129 (1989)
4. G. Ungar, Polymer **34**, 2050 (1993)
5. D.Y. Tsvankin, M.V. Gerasimov, V.P. Zhukov, I.I. Dubovik, D.R. Tur, V.S. Papkov, J. Polym. Sci., Part B: Polym. Phys. **30**, 851 (1992)
6. M.V. Gerasimov, *Phase transitions and structure of ordered phases in partly fluorinated polydialkoxyphosphazenes*, Ph. D. Thesis (in Russian) (Moscow, INEOS RAS, 1994)
7. L.L. Mack, R.J. Fitzpatrick, H.R. Allcock, Langmuir **13**, 2123 (1997)
8. A.I. Buzin, G. Brezesinski, D.R. Tur, V.S. Papkov, A.V. Bakirov, S.N. Chvalun, Macromolecules **48**, 3327 (2015)
9. J. Als-Nielsen, D. Jacquemain, K. Kjær, L. Franck, M. Lahav, L. Leiserowitz, Phys. Rep. **246**, 251 (1994)
10. R. Frahm, J. Weigelt, G. Meyer, G. Materlik, Rev. Sci. Instrum. **66**, 1677 (1995)
11. K. Kjær, Phys. B **198**, 100 (1994)
12. S.I. Belousov, A.I. Buzin, Y.K. Godovskii, Vysokomol. Soedin., Ser. B **41**, 1687 (1999)
13. T.R. Jensen, K. Kjær, G. Brezesinski, J. Ruiz-Garcia, H. Möhwald, N.N. Makarova, Y.K. Godovsky, Macromolecules **36**, 7236 (2003)
14. Y.K. Godovsky, G. Brezesinski, J. Ruiz-Garcia, H. Möhwald, T.R. Jensen, K. Kjær, N.N. Makarova, Macromolecules **37**, 4872 (2004)

First observation of 3D aggregates in a single-component Langmuir film below the equilibrium spreading pressure

George R. Ivanov[1,2] and Zorica Polevska[2]

[1]NanoBioSense Ltd., compl. Yavorov, bl. 38, entr. B, 1111 Sofia, Bulgaria
[2]University Laboratory for Nanoscience and Nanotechnology, University of Architecture, Civil Engineering and Geodesy, blvd. Hr. Smirnenski 1, 1046 Sofia, Bulgaria

Abstract. Single component monolayers from Dipalamitoyl Phosphatidyl Ethanolamine head labelled with the fluorescent chromophore NitroBenzoxaDiazole (DP-NBD-PE) were investigated at the air-water interface as Langmuir films and deposited on silicon wafers or glass plates as Langmuir-Blodgett (LB) films. A step compression and monitoring of the pressure relaxation together with domain formation as observed with Brewster angle microscopy suggests that main transition from liquid expanded to liquid compressed state start at around 6.9 mN/m though on the isotherm it starts around 9 mN/m at 25° C. Brewster angle microscopy also reveals a nonuniform structure in the monolayer. 3D aggregates - cylinders with 50 – 150 nm diameter and bilayer height were observed with Atomic Force Microscopy when deposition was carried at 7 mN/m, above the main phase transition but considerably lower than the equilibrium spreading pressure of 19.6 mN/m for this molecule. When LB film deposition is carried just below the main phase transition a uniform height layer of film in a liquid phase is observed with AFM with no structures. Both compression and deposition were carried at very low speeds with large time to relax in ordered to avoid kinetic effects. These 3D aggregates are not due to the transfer process or interaction with the substrate. These aggregates provide a highly developed area combined with monolayer thick structure which can produce very fast and highly sensitive biosensors.

1 Introduction

Langmuir-Blodgett film field has matured in the past 30 years. However systematic investigation of a structurally complex molecules still allows discovery of new effects. We have started the investigation of fluorescently labelled phospholipids some 28 years ago and over the years we have discovered 3 new effects [1, 2]. We have focused our efforts on fluorescently head labelled phospholipid Dipalmitoyl-Nitrobenzoxadiazole-Phosphatidyl Ethanolamine (DP-NBD-PE). Our focus in such type of compound was due to 2 reasons: the biomimetic aspect of a research of a molecule which is similar to the lipids in biological membranes and understanding of these more complex molecules. NBD chromophore is considered most suitable for optical biosensors due to high sensitivity to the environment in close proximity [3].

Here we elaborate on recently discovered new effect – formation of stable 3D nanosized aggregates at pressures considerably lower than the equilibrium spreading pressure (ESP). In previous research there was a considerable debate on 3D aggregates. But these structures were observed above the ESP in which the monolayer is in a metastable state. Or it was a mixture of several compounds. Instantaneous nucleation was proposed as a mechanism of collapse [4] but at high, collapse surface pressures. Sequential collapse in Langmuir films was also observed starting in the solid phase region [5] but no information on the ESP was provided in this case. In most cases, the π–A isotherms of amphiphiles are in a metastable state above the ESP and are in a state of supersaturation. This region of supersaturation is confined by ESP and the "collapse pressure" at which the irreversible monolayer collapse starts. The main characteristics of the supersaturated state is that the monolayers held at constant area per molecule or constant surface pressure show relaxation phenomena discussed in the terms of slow collapse.

2 Materials and methods

1,2-dipalmitoyl-sn-glycero-3-phosphoethanolamine-N-(7-nitro-2-1,3-benzoxadiazol-4-yl) (DP-NBD-PE) (figure 1) in chlorophorm solution was purchased from Avanti Polar Lipids. The claimed purity was above 99% and was used without further purification. This is a head labelled phospholipid with two saturated 16 carbon atoms tales. For

* Corresponding author: george@at-equipment.com

DOI: 10.1051/matecconf/20179801003

the Langmuir film data presented here a newly ordered and just opened compound was used.

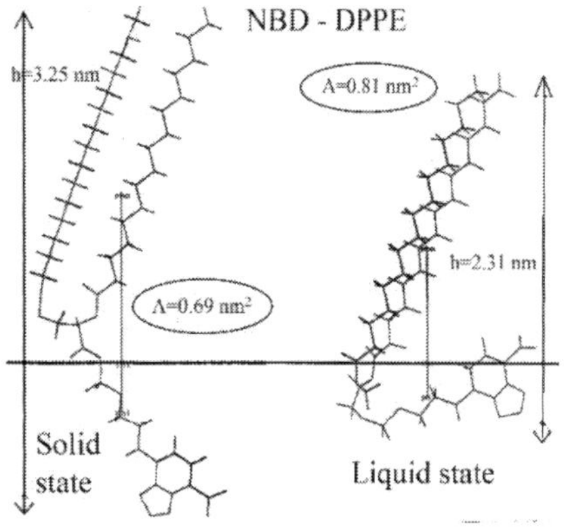

Fig.1 Structure of the DP-NBD-PE molecule and its most probable conformations according to molecular conformational analysis at simulated air-water interface in solid and liquid state.

π–A isotherms were measured on a long trough from KSV-NIMA (Finland). Together with the isotherms surface potential and Brewster angle microscopy (BAM) were measured, both manufactured from the same company. For the experiment shown a compression in steps of 2-4 mN/m was performed and then the equilibration of the system was monitored over time. For the AFM data a very slow constant velocity compression was used prior to the LB deposition which was also carried at extremely low withdrawal velocity.

For the Atomic Force Microscopy (AFM) an LB monolayer film was deposited on the upstroke on Si wafer substrate at different surface pressures. Si wafer was used because it has less surface roughness. Anfatec (Germany) Level AFM was used in dynamic mode.

3 Results and discussion

Surface pressure – mean molecular area (π – A) isotherms at different temperatures on pure water or in the presence of different metal counter ions was presented earlier [6]. Measurements were combined with fluorescence microscopy of the Langmuir film. The main phase transition from liquid expanded to liquid condensed phase at 20 ° C on pure water was at 6.9 mN/m. On figure 2 a more precise measurement of the isotherm is shown with compression in steps. After a step the film was maintained at constant surface pressure and monolayer relaxation was monitored through barrier position movement over time. Surface potential was measured simultaneously and BAM images were taken. They are shown in figures 2, 3 and 4 at increasing surface pressures.

At first look the main phase transition starts at 25 °C at surface pressure around 9 mN/m with a typical almost horizontal plateau. But looking at the barrier position vs time data (not shown) one can see that the phase transition starts below 6 mN/m. When we maintained constant surface pressure at 6 mN/m which is evidenced with the horizontal region on figure 2 the average barrier speed increases with time indicating that some molecules transfer from liquid to solid conformation. Formation of solid domains at this pressure is observed with BAM (figure 3).

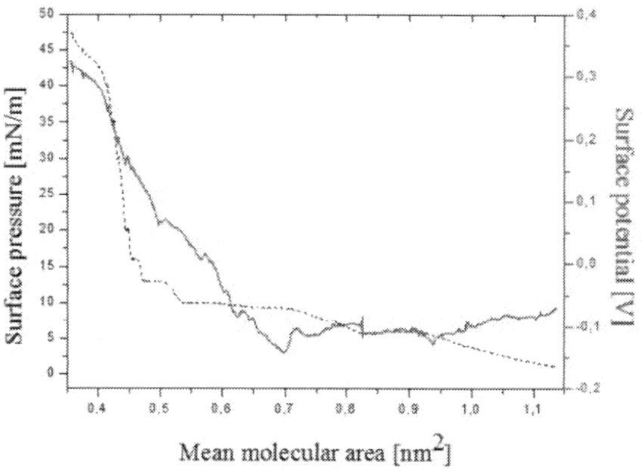

Fig. 2. Surface pressure (dotted line) and Surface potential as function of mean molecular area for a Langmuir film from DP-NBD-PE on pure water at 25° C.

In the middle of the main phase transition at mean molecular areas (MMA) around 0.65 nm² the surface potential starts to increase (figure 2). At 20° C the extrapolated to zero surface pressure of solid phase was 0.66 nm² which is in excellent agreement with our molecular conformational analysis at simulated air-water interface which predicts a MMA of the solid phase of 0.69 nm² [6]. The main absorption band at 460 nm for NBD-labelled lipids is due to intramolecular charge transfer [7], which is accompanied by a large (~ 4 Debye) change in dipole moment [8]. As the experiments here were carried in light this dipole moment is present. Molecular conformational analysis at simulated air-water interface suggests that in liquid phase the head group region is horizontal, even pointing a bit up towards the air-water interface while in the solid phase it point almost vertically down towards water bulk (see figure 1). There is almost 90° change in the dipole moment orientation which gives excellent contrast when scanning surface potential microscopy was used [9]. This middle region of the plato of the main phase transition corresponds to figure 4. Circular domains from the liquid compressed phase have formed. The domains are with different diameters, suggesting a second generation of crystallization centres. On a higher resolution inspection with fluorescence microscopy [10] and AFM [6] the domains are initially almost circular but at higher pressure become dendridic. This shape change is due to a balance between the line surface tension which tries to keep domains circular minimizing their perimeter and the repulsive forces between similarly oriented molecular dipoles which tries to separate molecules further apart. On a higher resolution AFM of a DP-NBD-PE LB films the solid domains do not have a strict boundary line but are composed of "archipelagos" and small islands floating in the liquid phase "sea".

At higher surface pressures the solid domains get in touch with each other (figure 5) and the slope of the π – A isotherm increases. The surface potential continues to grow almost linearly due to a larger number of solid domains entering the area below the vibrating electrode.

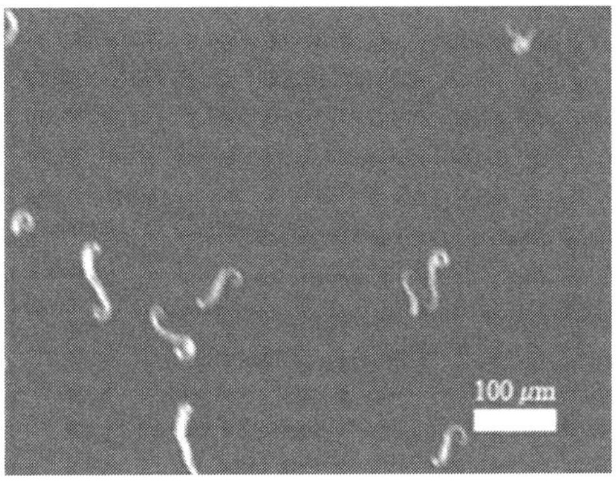

Fig. 3. BAM from a DP-NBD-PE Langmuir film at surface pressure 5.9 mN/m and mean molecular area of 0.82 nm^2.

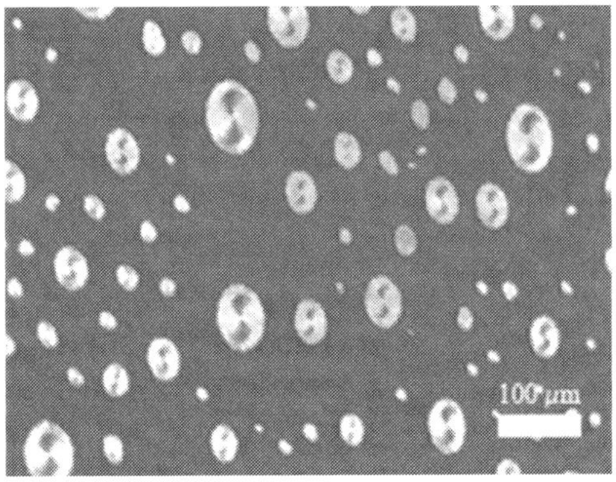

Fig. 4. BAM from a DP-NBD-PE Langmuir film at surface pressure 9.35 mN/m and mean molecular area of 0.65 nm^2.

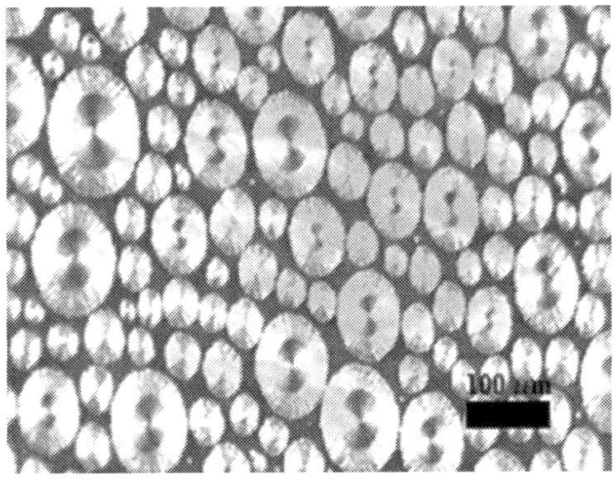

Fig. 5. BAM from a DP-NBD-PE Langmuir film at surface pressure 13 mN/m and mean molecular area of 0.47 nm^2.

On figure 5 the structure of the solid domain is clearly seen. It is composed of a central circular region with 2 pie type sections with different molecular tilt orientations. Around this central region is another region composed of smaller stripes with not so high polarization contrast. Initially domains are circular growing from a single nucleation centre. Afterwards at increasing surface pressure smaller domains start to attach to the bigger central domain [10]. This mechanism was observed also in a Langmuir film from DP-NBD-PE at 5° C [6] where higher surface pressure was needed to overcome the repulsion between the domains and they start to merge.

On figure 6 the formation of 3D nanosized aggregates in a single component LB film deposited significantly below the ESP which for this molecule was measured to be 19.6 mN/m [6] is shown. When LB film deposition is carried below the main phase transition a uniform liquid phase is observed with AFM. However just above the beginning of the main phase transition at around 7 mN/m cylindrical structures with bilayer height are observed. If the deposition is carried out at higher than the ESP surface pressures these cylinders grow in height initially up to 13 nm for deposition at 33 mN/m and up to 35 nm, and in some cases to hundred nanometers at 43 mN/m. However, if the monolayer is allowed to relax under normal laboratory conditions for some time (50 days in this case), those cylinders again become with bilayer height [6]. Typical diameter of the cylinders is 50 to 150 nm.

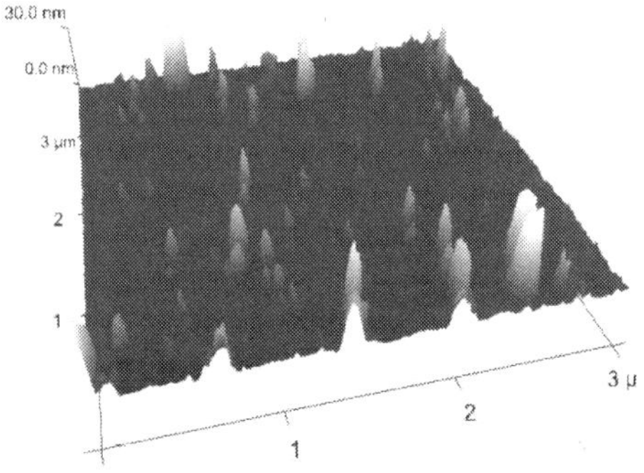

Fig. 6. AFM of an LB film monolayer from DP-NBD-PE transferred at 10 mN/m, room temperature and pure water.

These cylinders can be due to several artefacts. Contamination can be excluded because the substance is more than 99 % pure and all other typical precautions were taken. Kinetic effects can also be eliminated because there is a complete match between the fluorescence microscopy pictures of the Langmuir film and the LB monolayer on the substrate [11]. The biggest artefact problem can be whether there are no alterations during the transfer process from the air-water interface to the solid substrate, i.e. whether these 3D cylinders are not a result of the LB film deposition or an interaction with the substrate. It seems that this depends on the substance transferred. H. Riegler used a tilted substrate and a fluorescence microscope to monitor the lipid film deposition with 2 % label [12]. With the substrate stationary, upon increasing the surface pressure, substrate mediated condensation of probe-depleted stripes was observed. During the slow Langmuir-Blodgett transfer (i.e. upstroke of the substrate under constant surface pressure), a domain-free gap is observed adjacent to the three-phase line. Both phenomena were caused by the substrate mediated condensation of the monolayer onto the solid surface in the moment of its deposition. On the other hand D. Vollhardt et al. investigating 1-monostearoylglycerol monolayers and comparing BAM and friction mode AFM data suggests that after the deposition of the monolayer, the main features of the condensed phase morphology, such as size, shape, and azimuthal tilt, are preserved independent of the solid substrate [13]. Similar shape and size of the condensed phase domains for two different systems were also found [14]. So it seems highly unlikely that these 3D aggregates in our experiment are formed during the deposition process. Further on they were observed on quite large number of experiments and depositions both on Si wafer or glass substrates.

We have not measured directly the structure of these 3D cylinders but some integral measurements can help. Data from our Stark spectroscopy measurements show that DP-NBD-PE molecules tend to form centrosymmetrical non-polar structures. Thus for the second layer in the bilayer structures the molecules most probably flip over and we have a tail – tail contact. Comparing fluorescence lifetime imaging microscopy data with integrally measured fluorescence lifetime we can suggest that the molecules in the 3D cylinders are in the liquid phase with much higher fluorescence lifetime compared to the solid phase [15].

4 Conclusion

The combination of $\pi - A$ isotherm with step compression and monitoring of the relaxation processes, surface potential, BAM and AFM measurements gives detailed picture of the behaviour of DP-NBD-PE single component monolayers both as Langmuir and LB films. Current measurements are combined with our systematic previous research on this compound using almost all possible experimental methods and computer-aided molecular conformational analysis at simulated air-water interface. When the LB film deposition was carried above 7 mN/m at 20^{0} C where the liquid expanded to liquid condensed phase transition begins, 3D aggregates with bilayer height and 50 to 150 nm diameters are observed. This deposition pressure is significantly below the ESP suggesting that these are not some metastable effects. These cylinders are stable over time at least on a months scale.

Acknowledgments

This work is part of NanoBioSense Ltd. development program. It was also supported by a grant number 176/15 with the Centre for Design and Innovation of the University of Architecture, Civil Engineering and Geodesy, Sofia, Bulgaria. Invitation to visit and perform some measurements in the laboratory of KSV-NIMA from Biolin Scientific group is gratefully acknowledged. Special thanks to Jyrki Korpela from KSV-NIMA for the help with the measurements.

References

1. G.R. Ivanov, "First observation of fluorescence self-quenching in Langmuir films", Chem. Phys. Lett., **193** 323 (1992)
2. G.R. Ivanov and M. Fujihira, "Novel phenomena in organized organic monolayers", in *Thin Film Materials and Devices - Developments in Science and Technology*, eds. J. M. Marshall, N. Kirov, A. Vavrek, J. M. Maud, World Scientific, Singapore, 261 (1999)
3. J.D. Brennan, K.M.R. Kallury and U.J. Krull, Thin Solid Films, **244**, 898 (1994)
4. D. Vollhardt, Colloids and Surfaces A, **156**, 79 (1999)
5. E. Hatta, Langmuir, **20**, 4059 (2004)
6. G.R. Ivanov, G. Georgiev and Z. Lalchev, Fluorescently Labeled Phospholipids – New Class of Materials for Chemical Sensors for Environmental Monitoring, in *Relevant Perspectives in Global Environmental Change,* ed. J. Agboola, ISBN: 978-953-307-709-3, Rijeka: (InTech) 89 (2011)
7. P.A. Paprica, N.C. Baird and N.O. Petersen. J. Photochem. Photobiol. A: Chem., **70**, 51 (1993)
8. S.A. Mukherjee, A. Chattopadhyay, A. Samanta and T. Soujanya, Phys. Chem., **98**, 2809 (1994)
9. G.R. Ivanov and M. Fujihira, "Single component Langmuir-Blodgett film investigated with 5 scanning probe microscopy methods", Colloids and Surfaces A, **198 – 200**, 305 (2002)
10. G.R. Ivanov and N.I. Geshev, „New effects in Langmuir and Langmuir-Blodgett monolayers from fluorescently labelled phospholipids and their possible use for water quality control", IOP Science, *Journal of Physics: Conference Series*, **682**, 012037 (2016)
11. G.R. Ivanov, J. Nanoscience and Nanoengineering, **2**, 15 (2016).
12. H. Riegler and K. Spratte, Thin Solid Films, **210 – 211**, 9 (1992)
13. K. Meine, D. Vollhardt and G. Weidemann, Langmuir, **14**, 1815 (1998).
14. L. Santesson, T.M.H. Wong, M. Taborelli, P. Descouts, M. Liley, C. Duschl, H. Vogel, J. Phys. Chem. **99**, 1038 (1995).
15. G.R. Ivanov and J. Burov, AIP Conf. Proc. **899**, 257 (2007)

Systematic study of Langmuir films of different amino acid derivatives on several subphases

L. Románszki[1,*] and *J.* Telegdi[1,2]

[1] Research Centre for Natural Sciences, Hungarian Academy of Sciences, Institute of Materials and Environmental Chemistry, 1117 Budapest, Magyar tudósok körútja 2., Hungary
[2] Óbuda University, Faculty of Light Industry and Environmental Engineering, 1034 Budapest, Doberdó u. 6., Hungary

Abstract. In a Langmuir trough, we studied the formation and the behaviour upon compression of Langmuir films of 1) α-hexadecyl derivative of *tert*-butoxycarbonylated glycine, respectively α-octadecyl derivative of *tert*-butoxycarbonylated glycine methyl ester on L- and D-amino acid subphases with/without Cu^{2+} ions, and 2) *N*-hexadecanoyl derivatives of four L-amino acids (Ala, Ser, Glu, and Phe) on pure water subphase. The films were characterized by their A_c and A_0 molecular areas, Π_c collapse pressure, C^{-1} compression modulus, and the $\Pi_{C^{-1}_{max}}$ surface pressure at maximum compression modulus. In this contribution, the found peculiarities of the different amphiphile–subphase pairs and the drawn conclusions about the structure of the films and the interactions of the amphiphiles with the subphases are presented.

1 Introduction

Langmuir (L) films and Langmuir–Blodgett (LB) layers of long side chain amino acid derivatives are interesting for the investigation of (bio)sensing mechanisms, drug delivery processes, 2D chiral organization and recognition, etc., and the development of sensor, corrosion inhibitor, anti-biofouling etc. layers. L films of Ala [1–3], Asp [4], Ser [5–6], Val [3] etc. alkyl derivatives have been studied on different subphases.

In our recent studies, we demonstrated the corrosion protection and antibiofouling ability of carboxylic and hydroxamic acid LB layers and self-assembled monolayers [7–12]. In the present study, the L films of two α-alkyl amino acid derivatives and four *N*-hexadecanoyl amino acids on several subphases have been studied by surface pressure – area isotherms.

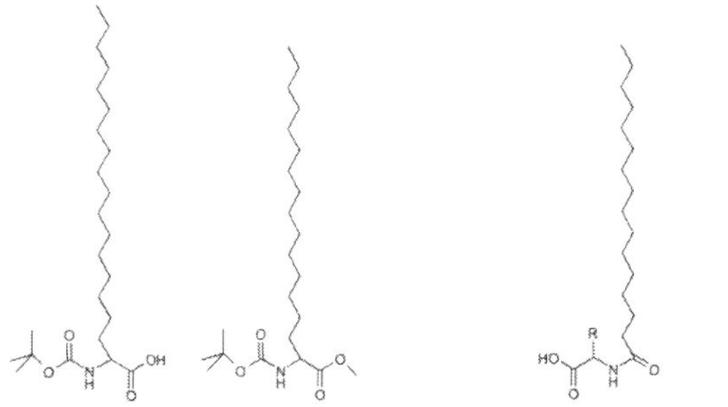

Fig. 1. Amphiphilic compounds used in these experiments: α-alkyl amino acid derivatives BOC–Gly–C18 and BOC–GlyMe–C16, respectively *N*-palmitoyl amino acids C16–*N*–Ala, C16–*N*–Ser, C16–*N*–Glu and C16–*N*–Phe.

2 Materials and methods

The raceme mixtures of the two α-alkyl amino acid derivatives, *N*-(*t*-butoxycarbonyl)-2-hexadecyl-glycine methyl ester (BOC–GlyMe–C16) and *N*-(*t*-butoxy-carbonyl)-2-octadecyl-glycine (BOC–Gly–C18), (see figure 1) have been synthesized as described in [13]. L-*N*-hexadecanoyl amino acids were prepared in a Schotten–Baumann reaction from the respective L-amino acid and hexadecanoyl chloride. The resulting compounds were purified in subsequent recrystallization steps, and were characterized by elemental analysis, melting point determination, thin layer chromatography and optical rotation. 30, 45 and 60 µl volumes of 2 mM solutions in chloroform were uniformly spread in a LB trough (611D, NIMA Technology Ltd., Coventry, UK) equipped with a Wilhelmy-type surface pressure gauge using filter paper. The *N*-hexadecanoyl amino acids were spread on a clear water subphase (MilliQ). The α-alkyl amino acid derivatives were spread on: water, 1 mM $CuAc_2$, 100 mM D-Ala, 100 mM L-Ala, 100 mM D-Asn, 100 mM L-Asn, and mixed solutions containing 100 mM D-Asn + 1 mM $CuAc_2$, respectively 100 mM L-Asn + 1 mM $CuAc_2$.

* Corresponding author: romanszki.lorand@ttk.mta.hu

Compression of the monolayers started 2 minutes after spreading, with a rate of 100 cm²/min. The subphases were thermostated to 20 °C with a precision of 0.5 °C. The obtained surface pressure – area isotherms were further processed to obtain collapse pressure, molecular areas and compression moduli.

3 Results and discussion

3.1 α-alkyl amino acid derivatives

The isotherms of BOC–GlyMe–C16 and BOC–Gly–C18 on different subphases are shown in figure 2, A and B. The derived compression modulus curves are presented in figure 2, C and D. BOC–GlyMe–C16 shows a continuous increase of the surface pressure, going through an inflexion point, and then increasing with a mild slope without a breakdown. Contrarily, the isotherms of BOC–Gly–C18 initially start to increase with a moderate slope, go through a plateau region, and then rise steeply and finally a real collapse is observed.

The BAM images of the two amphiphile's monolayers (not shown) are accordingly very different. In the case of the BOC–GlyMe–C16 individual islands formed already at the moment of spreading with clear boundaries and high contrast can observed, whereas BOC–Gly–C18 spreads instantaneously and homogeneously on the surfaces forming continuous monolayers.

From the isotherms, the A_c molecular area at collapse and the corresponding Π_c collapse pressure are obtained. By extrapolation of the collapse point to zero surface pressure, the A_0 molecular area can be determined. Table 1 and figure 3 summarize these quantities. On each subphase 1) the collapse pressures of the BOC–Gly–C18 films are much higher; and 2) both molecular area values of the BOC–Gly–C18 films are much smaller than those of the BOC–GlyMe–C16 films.

The BOC–GlyMe–C16 monolayers collapse before reaching a true solid phase, continuously, the film layering up as the barrier moves. This collapse is very similar to the quasistatic collapse found at unsaturated fatty carboxylic and hydroxamic acid monolayers [7]. In contrast, the BOC–Gly–C18 films collapse instantaneously, after reaching a real solid phase.

These observations indicate that BOC–Gly–C18 molecules behave much more ideally than BOC–GlyMe–C16 molecules. The BOC–Gly–C18 molecules interact with the water molecules through their free carboxylic groups. This interaction is apparently much stronger than the one among the film forming molecules themselves. Thus, the molecules spread readily and quickly on the water surface, covering the whole area uniformly. Contrarily, BOC–GlyMe–C16 molecules behave non-ideally. Probably due to the esterified carboxyl group, which is more hydrophobic than a free one, the molecules adopt distorted conformations with much larger space demand. They also tend to form aggregates from the very beginning of the spreading, the

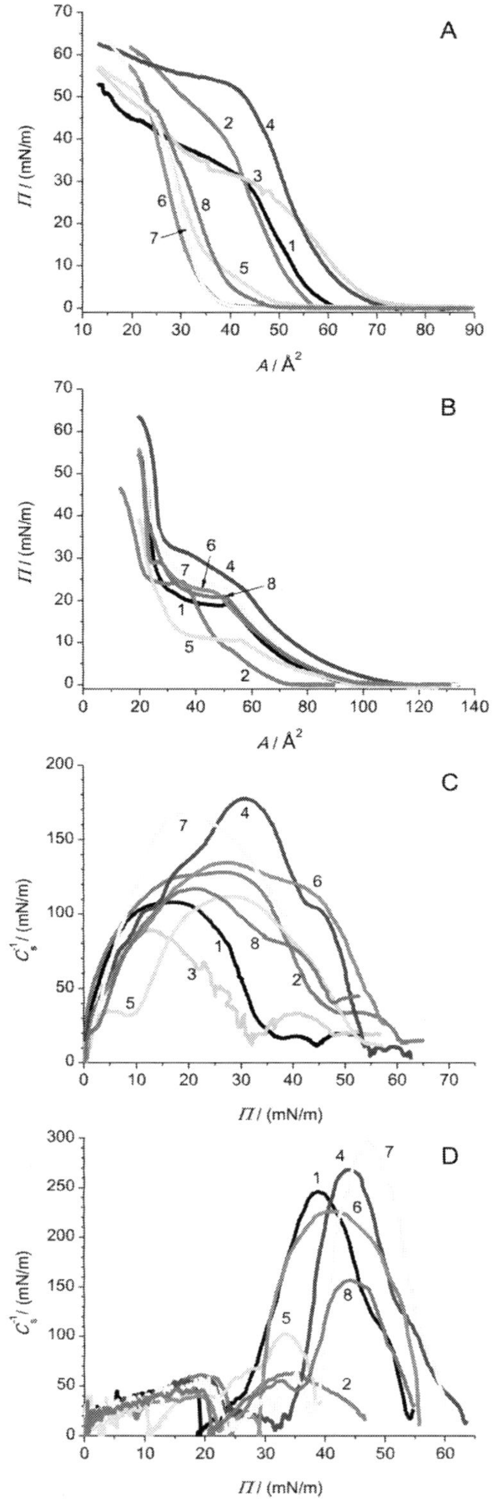

Fig. 2. Surface pressure (Π) – molecular area (A) isotherms of BOC–GlyMe–C16 (A) and BOC–Gly–C18 (B), and the corresponding compression modulus (C_S^{-1}) – surface pressure (Π) curves (C and D) at 20 °C on different subphases: 1) water; 2) CuAc₂; 3) D-Ala; 4) L-Ala; 5) D-Asn; 6) D-Asn + CuAc₂; 7) L-Asn; 8) L-Asn + CuAc₂.

interaction among the amphiphiles being stronger than the interaction between the amphiphiles and the water.

Table 1. Π_c collapse pressures as well as A_c and A_0 molecular areas of L monolayers of BOC–GlyMe–C16 and BOC–Gly–C18 on different subphases. For each amphiphilic compound, the subphases with lowest and highest values are marked with light, respectively dark grey.

# subphase	BOC–GlyMe–C16			BOC–Gly–C18		
	Π_c mN/m	A_c Å²	A_0 Å²	Π_c mN/m	A_c Å²	A_0 Å²
1 water	19.3	48.3	50.8	38.0	23.9	27.1
2 CuAc2	28.3	43.4	48.5	30.9	19.7	24.4
3 D-Ala	27.3	48.4	53.0			
4 L-Ala	29.7	51.0	57.1	42.4	26.3	29.1
5 D-Asn	31.6	28.8	35.4	32.6	21.9	27.1
6 D-Asn + CuAc2	28.1	27.6	32.1	42.7	22.0	25.4
7 L-Asn	27.0	29.4	33.8	44.8	23.2	26.0
8 L-Asn + CuAc2	21.6	33.6	37.8	44.3	22.2	27.4

Metal ions, as well as certain amino acids added to the subphase, are known to interact with the film forming molecules through strong coordinative and hydrogen bonds. The change in Π_c, A_c and A_0 of these films with respect to those on pure water subphase, can be used to conclude about the strength and geometry of these bonds.

The copper ions in the subphase have, as expected, a compacting effect on the organization of the amphiphiles, decreasing both A_c and A_0. This tendency is more pronounced in the case of BOC–Gly–C18 than in the case of BOC–GlyMe–C16 that can be attributed to the presence of the methyl ester group in BOC–GlyMe–C16, as described above.

On both D- and L-Ala subphases, BOC–GlyMe–C16 monolayers exhibit slightly larger A_c and larger A_0 values than on pure water. Similarly, BOC–Gly–C18 monolayer on L-Ala subphase exhibits larger A_c and A_0 than on pure water. The Π_c collapse pressures are in each case higher on these amino acid subphases than on pure water subphases, indicating stiffer films. Thus Ala, in its both optical isomers, slightly expands the monolayers of these amphiphiles.

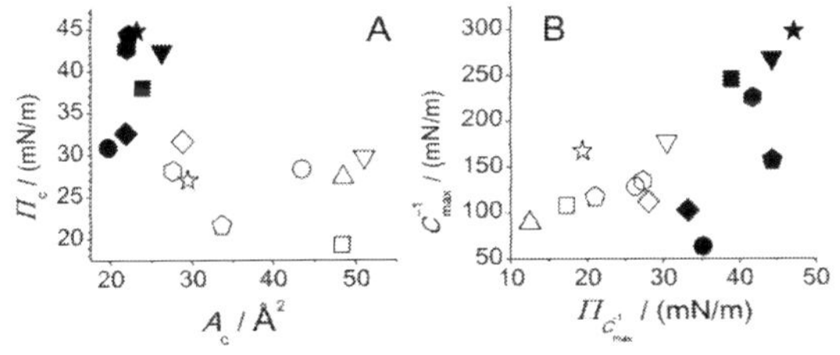

Fig. 3. Π_c collapse pressure and A_c molecular area values (A), respectively C^{-1}_{max} maximum compression moduli and the corresponding $\Pi_{C^{-1}_{max}}$ surface pressure values (B) determined from the L film isotherms of BOC-GlyMe-C16 (empty symbols), respectively BOC-Gly-C18 (filled symbols), recorded on different subphases: water (■); CuAc2 (●); D-Ala (▲); L-Ala (▼); D-Asn (◆); D-Asn + CuAc2 (●); L-Asn (★); L-Asn + CuAc2 (⬟).

D-Asn, contrarily, has a close-packing effect: both BOC–GlyMe–C16 and BOC–Gly–C18 exhibit smaller molecular areas on D-Asn than on pure water. This packing effect is more pronounced in the case of BOC–GlyMe–C16 than in the case of BOC–Gly–C18. For BOC–GlyMe–C16, it is even stronger than the effect of copper ions. By adding copper ions to the D-Asn subphase, the molecular areas decrease slightly even further. The effect of L-Asn is qualitatively similar and quantitatively very close to that of D-Asn: for BOC–GlyMe–C16, A_c and A_0 are 39%, respectively 33% smaller than on water, whereas for BOC–Gly–C18, by only 3%, respectively 4%. Interestingly, by adding copper ions to the L-Asn subphase, the molecular areas of the BOC–GlyMe–C16 monolayers decrease somewhat less as compared to the values on pure water. However, in the case of BOC–Gly–C18, A_c decreases slightly more on L-Asn + CuAc2 than on the pure L-Asn subphase, while A_0 practically does not change.

3.2 *N*-hexadecanoyl amino acids

The isotherms and the derived compression modulus curves of the *N*-hexadecanoyl amino acid L films on pure water subphase are presented in figure 4. Figure 5 shows the determined Π_c collapse pressures (A), A_c (B) and A_0 (C) molecular areas and the difference of these (D). In figure 6, the position of the amphiphiles in a Π_c–A_c coordinate system is presented.

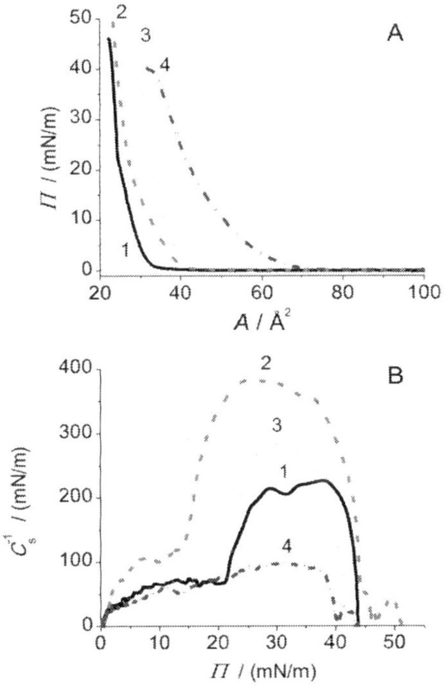

Fig. 4. Typical Π–A isotherms (A) and the corresponding C_s^{-1}–Π curves (B) of the Ala (1), Ser (2), Glu (3) and Phe (4) C16-amino acids.

The L films of the Ala and Ser derivatives exhibit almost identical molecular areas and collapse pressures, most likely due to the similarities in the molecular structures. The molecular areas of the Glu derivative L film are larger, with lower collapse pressure. The largest molecular area and the lowest collapse pressure were found in the case of the Phe derivative.

These results show strong correlation with the size of

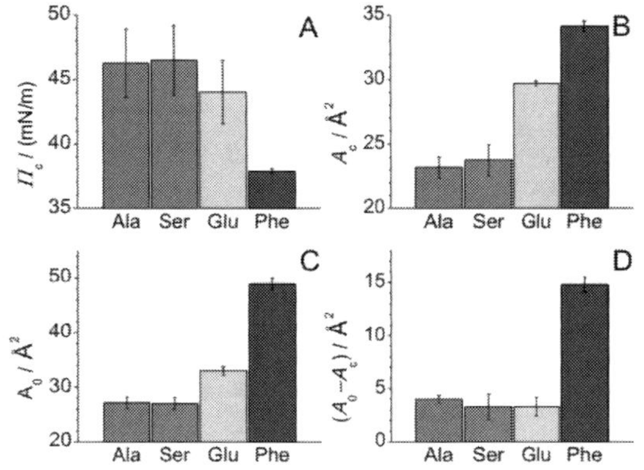

Fig. 5. Π_c collapse pressures (A), A_c molecular areas at collapse (B), A_0 molecular areas extrapolated to zero surface pressure (C) and the difference of the latter two (D) of the C16-amino acid derivatives L films.

the base amino acid (Ala ≈ Ser < Glu < Phe): with increasing amino acid size, the packing density of the L films is decreasing. Therefore it is expected that the most suitable compounds for densely packed, well ordered LB layer preparation, in about equal extent, are the C16–*N*–Ala and C16–*N*–Ser, due to the relatively small size of the amino acid part, and additionally, in the case of C16–*N*–Ser, the hydrophilicity of the OH group. Contrarily, the large and hydrophobic phenyl group of C16–*N*–Phe prevents molecules from tight packing, and its hydrophobicity does not favour an interaction with the water subphase.

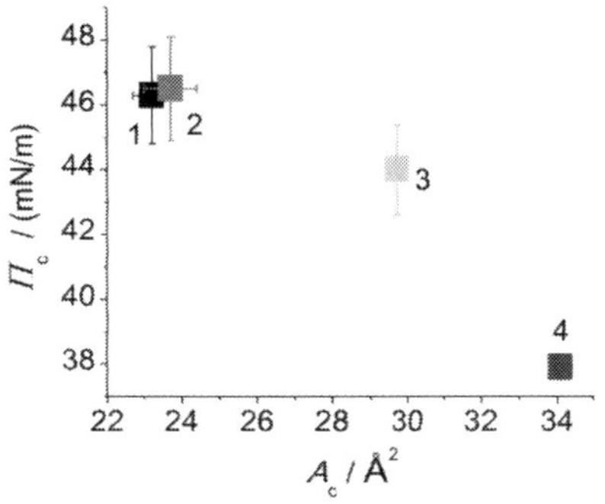

Fig. 6. The place of the Ala (1), Ser (2), Glu (3) and Phe (4) C16-amino acid derivative L films in a Π_c–A_c coordinate system.

4 Conclusions

The presence or absence of a methyl ester group in the structure of the two α-alkyl amino acid derivative amphiphiles causes significant differences in the organization of the monolayers at the subphase/air interface. Cu^{2+} ions in the water subphase have a close-packing effect on the monolayers of α-alkyl amino acids. Different chiral amino acids in the subphase are interacting with the film-forming α-alkyl amino acid molecules at the interface. The extent of this interaction depends on the interacting partners and needs further investigation. By selecting the adequate subphase, well ordered, closely packed LB layers can be prepared. The properties of the L films of *N*-hexadecanoyl amino acids correlate with the size of the base amino acids. These results serve as good starting point for further investigations to clarify the role of the subphase and the strength of interaction between the L film and the compounds in the subphase, as well as the effect of temperature and pH. As the amino acids play important roles in transmembrane protein channels, we hope these answers would bring us closer to the development of new biosensing mechanisms.

Acknowledgement

The authors thank Katalin Tímár for the synthesis of the *N*-alkanoyl amino acid derivatives.

References

1. O. Bouloussa, M. Dupeyrat, Biochim. Biophys. Acta **938**, 395 (1988)
2. F. Hoffmann, H. Hühnerfuss, K.J. Stine, Langmuir **14**, 4525 (1998)
3. F. Hoffmann, K.J. Stine, H. Hühnerfuss, J. Phys. Chem. B **109**, 240 (2005)
4. K. Thirumoorthy, N. Nandi, D. Vollhardt, Colloid. Surface. A **282–283**, 222 (2006)
5. N.G. Harvey, D. Mirajovsky, P.L. Rose, R. Verbiar, E.M. Arnett, J. Am. Chem. Soc. **111**, 1115 (1989)
6. K.J. Stine, J.Y.J. Uang, S.D. Dingman, Langmuir **9**, 2112 (1993)
7. L. Románszki, J. Telegdi, E. Kálmán, Colloid. Surface. A **321**, 20 (2008)
8. L. Románszki, J. Telegdi, L. Nyikos, Stud. Univ. Babes–Bolyai Chem. **57**, 191 (2012)
9. J. Telegdi, T. Szabó, L. Románszki, M. Pávai, In: *Handbook of smart coatings for materials protection* (Ed. ASH Makhlouf, Woodhead Publishing Ltd, Cambridge, 2014) 135
10. J. Telegdi, L. Románszki, F. Al-Taher, É. Pfeifer, E. Kálmán, In: *17th International Corrosion Congress 2008: Corrosion Control in the Service of Society* (NACE International, Houston, USA, 2009) 2631
11. L. Románszki, J. Telegdi, E. Kálmán, In: *Proceedings of 6th International Conference of PhD students. Natural Sciences.* (University of Miskolc, Hungary, 2007) 97
12. J. Telegdi, F. Al-Taher, L. Románszki, E. Kálmán, In: *The 3rd International Conference on Technological Advances of Thin Films & Surface Coatings (THINFILMS 2006)* (Singapore, 2006) paper no. TCR640
13. W.A. Gibbons, R.A. Hughes, M. Charalambous, M. Christodoulou, A. Szeto, A.E. Aulabaugh, P. Mascagni, I. Toth, Liebigs Ann. Chem. **12**, 1175 (1990)

Infrared external reflection spectra of Janus gold nanoparticle monolayers prepared at the air–water interface

H. Sakai[1,*], *A.* Matsuzawa[1] and *T.* Kawai[2]

[1]National Institute of Technology, Oyama College, 771 Nakakuki, Ooaza, Oyama, Tochigi 323-0806, Japan
[2]Tokyo University of Science, Kagurazaka 1-3, Shinjuku-ku, Tokyo 162-8601, Japan

Abstract. Nanoparticle monolayers at the air–water interface have received broad attention because of their potential applications owing to the tunability of their surface properties. We prepared gold nanoparticle monolayers by adsorption of hydrophilized nanoparticles from solution onto Langmuir monolayers of amphiphilic organic molecules (polyion complexes of deuterated stearic acid and polyallylamine) at the air–water interface, followed by adsorption of dodecyltrimethylammonium bromide onto the underside of these particles. The infrared external reflection spectra of the monolayers recorded both during and after preparation showed that the adsorption of gold nanoparticles on the Langmuir monolayers and that of dodecyltrimethylammonium bromide on the nanoparticles occurred in several tens of minutes. The infrared spectra also showed that both stearic acid and dodecyltrimethylammonium bromide molecules were highly ordered on the gold nanoparticles at the air–water interface. These results suggested that this preparation procedure is a useful fabrication method for Janus particles.

1 Introduction

The construction of 2D arrays of nanoparticles (NPs) is of great importance owing to their potential applications in electronic, magnetic, and optical devices [1-3]. Gold nanoparticle (AuNP) monolayers at the air–water interface have recently attracted widespread attention because of their tunable lateral surface density and ability to form 3D assemblies via the Langmuir–Blodgett method [4-17]. Moreover, NP monolayers can easily form anisotropic structures at the air–water interface because they are vertically confined. However, AuNP monolayers prepared by adsorption of hydrophilized AuNPs from solution onto Langmuir monolayers of amphiphilic organic molecules at the air–water interface have not been examined in as much detail as those prepared by spreading highly hydrophobized NPs over the air–water interface. The preparation method involving hydrophilized AuNPs allows for the fabrication of Janus particles through the addition of different ligands to the underside of particles partially covered with molecules in the Langmuir monolayer. Moreover, this method has the advantage of using little organic solvent during the fabrication of Janus particles. To the best of our knowledge, few studies have used this method to fabricate Janus particles [18]. It is therefore important to determine the adsorption kinetics and the structural arrangement of the ligands on the nanoparticles prepared using this method.

In this study, we prepared AuNP monolayers covered with dodecyltrimethylammonium bromide (DTAB) and Langmuir monolayers of polyion complexes of deuterated stearic acid (d-STA) and polyallylamine (PAA) at the air–water interface (Figure 1). The infrared external reflection (IER) spectra of the monolayers were recorded in situ. The structural arrangement and configurations of the molecules was estimated from the intensity and wavenumber of the band peaks in the IER spectra. We also observed the adsorption of the molecules on the nanoparticles during the preparation process.

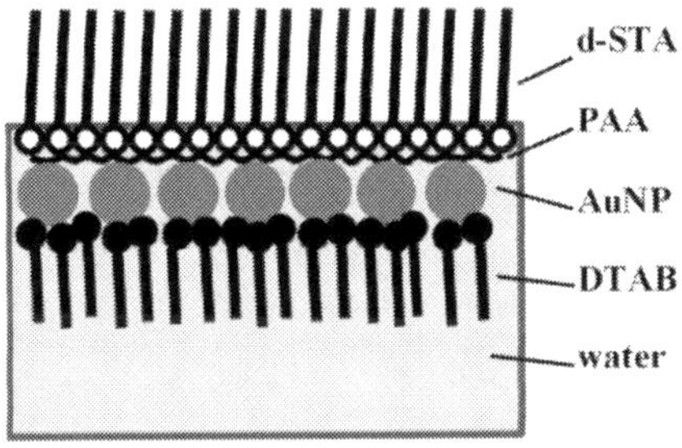

Fig. 1. Schematic illustration of AuNP monolayers covered with DTAB and Langmuir monolayers of d-STA and PAA.

* Corresponding author: sakai@oyama-ct.ac.jp

2 Materials and Methods

2.1 Materials

Hydrogen tetrachloroaurate(III) trihydrate (Sigma-Aldrich), trisodium citrate dehydrate (Cica), PAA (Nitto Boseki, MW: 8000), and d-STA (CDN ISOTOPES) were used without further purification. The deuterated compound was used so that we can differentiate between d-STA and DTAB in the IER spectra. DTAB (Sigma-Aldrich) was recrystallized from an acetone solution.

2.2 Preparation Method

AuNPs were prepared according to the standard sodium citrate reduction method [19]. Transmission electron microscopy (TEM) observations and dynamic light scattering measurements revealed that the mean diameter of the AuNPs was around 20 nm. The concentration of the AuNP solution was calculated to be 3.0×10^{15} particles/L.

AuNP monolayers at the air–water interface were prepared by adsorption of AuNPs onto Langmuir monolayers of polyion complexes of d-STA and PAA. The polyion complex was prepared in methanol at a molar ratio of 1:2 (d-STA:PAA, monomer units). A chloroform solution of the polyion complex was then spread onto the AuNP solution surface in a Langmuir trough to give a Langmuir monolayer with the surface area of 0.20 nm^2/d-STA molecule. After waiting 60 min to allow the AuNPs to adsorb onto the Langmuir monolayer, the AuNP solution was replaced with water as follows: 1 mL of the AuNP solution underneath the Langmuir monolayer was removed from the Langmuir trough (12 mL) using a pipette, and then 1 mL of water was added into the trough. This process was repeated 100 times. Subsequently, a DTAB solution was injected into the Langmuir trough. The concentration of DTAB was about 8 mM, which is less than the critical micelle concentration of DTAB. After waiting 60 min to allow DTAB molecules to adsorb onto the AuNP monolayer, the DTAB solution was replaced with water in the manner described above. The final concentration of DTAB in the trough was calculated to be about 0.001 mM.

2.3 IER

The IER spectra during and after the preparation of the NP monolayer were recorded at the air–water interface using a Nicolet iS50 FT-IR spectrophotometer equipped with a mercury–cadmium–telluride (MCT) detector with a resolution of 4 cm^{-1}. A trough with an 80×22 mm^2 effective surface area attached to a Specac 19650 monolayer/grazing angle accessory was used. The incident angle from the surface normal was 40°. An unpolarized beam was used, and 500 interferograms were collected. The vertical axis of the reflection spectrum was defined as $\log(R_0/R)$, where R and R_0 are the sample and background reflectance, respectively. The background reflectance was recorded on the surface of either AuNP solution or water, according to the subphase of the each sample.

2.4 TEM

AuNP monolayer samples for TEM measurements were fabricated on a 250 mesh carbon-covered Cu grid (Okenshoji ELS-C10) using a horizontal method [8]. The samples were examined using a JEOL JEM-1011 transmission electron microscope operating at 100 kV.

3 Results and Discussion

Figure 2(a) shows the TEM image of a TEM sample that was fabricated 60 min after spreading the polyion complex solution on the AuNP solution. The AuNPs formed a monolayer underneath the Langmuir monolayer of the polyion complex. Hereafter, we designate this monolayer as a d-STA/PAA/AuNP film. It has been reported that AuNPs interact with ionic molecules in Langmuir monolayers through ionic bonding to form monolayers of AuNPs [4, 6-9, 15, 17]. The present result indicates that ionic bonding likely occurred between the NH_3^+ and COO^- groups of PAA and AuNPs, respectively.

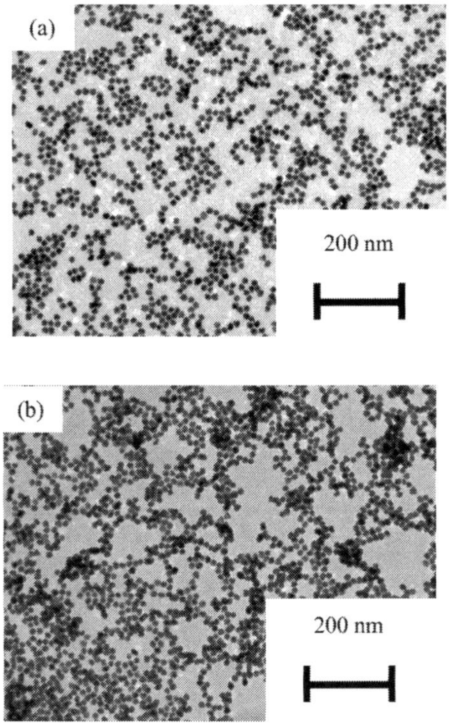

Fig. 2. TEM images: (a) STA/PAA/AuNP film before replacement of the AuNP solution with water and (b) d-STA/PAA/AuNP/DTAB film after replacement of the DTAB solution with water.

Figure 2(b) shows the TEM image of a TEM sample that was fabricated after replacing the DTAB solution with water. We designate this film as d-

STA/PAA/AuNP/DTAB. The film remained on the water surface, even after replacing the solution (AuNP and DTAB) with water twice. To the best of our knowledge, there have been few reports about AuNP monolayers subjected to such harsh treatments. The outstanding stability of the present film can probably be attributed to the strong ionic bonding interactions and the robustness of the polyion complex in the Langmuir monolayer.

Figure 3 shows the IER spectra of the d-STA/PAA/AuNP film on the AuNP solution and water in the CH_2 and CD_2 stretching vibrational region. The times shown in this figure represent the elapsed time after spreading the polyion complex solution on the surface of the AuNP solution. The peaks corresponding to the antisymmetric and symmetric CD_2 stretching band of d-STA are observed in the IER spectra. Figure 4 shows a plot of the intensity (peak height) of the antisymmetric CD_2 stretching band against the elapsed time on the AuNP solution and on the water surface. On the AuNP solution, the absorbance increases for several tens of minutes, probably owing to surface-enhanced infrared absorption [20] because of adsorption of AuNPs underneath the Langmuir monolayer. This result indicates that the d-STA/PAA/AuNP film was formed in several tens of minutes and then remained intact. The appearance of this band at 2193 cm^{-1} indicated that the conformation of the d-STA molecule was all-*trans* [21]. The intensity and wavenumber of this band remained largely unchanged after replacing the AuNP solution with water.

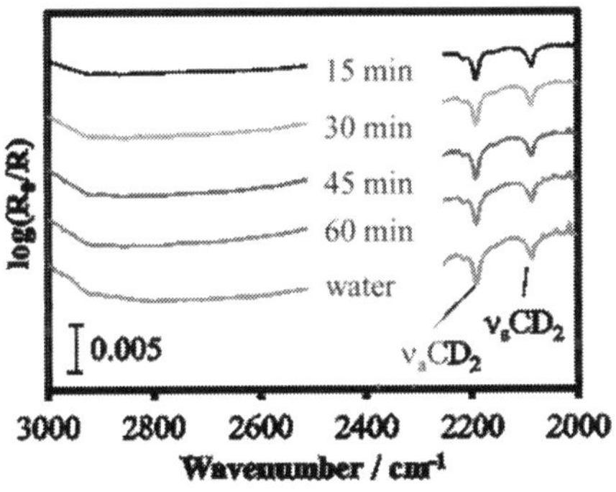

Fig. 3. IER spectra of the d-STA/PAA/AuNP film on the AuNP solution and water.

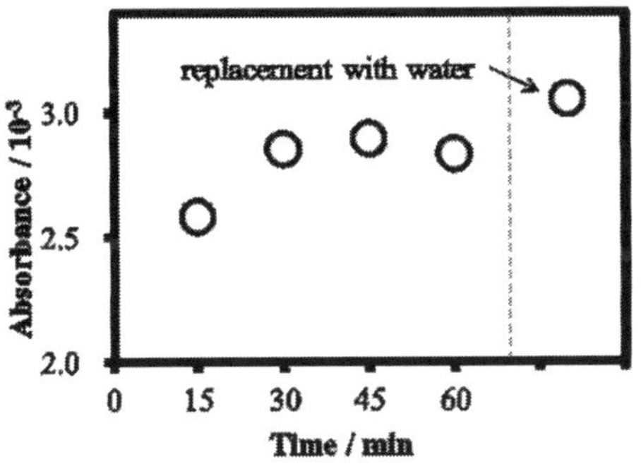

Fig. 4. v_aCD_2 absorbance vs. time on the AuNP solution and water.

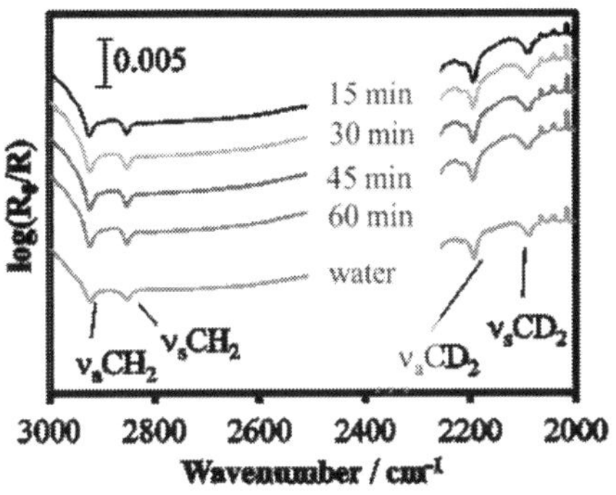

Fig. 5. IER spectra of the d-STA/PAA/AuNP/DTAB film on the DTAB solution and water.

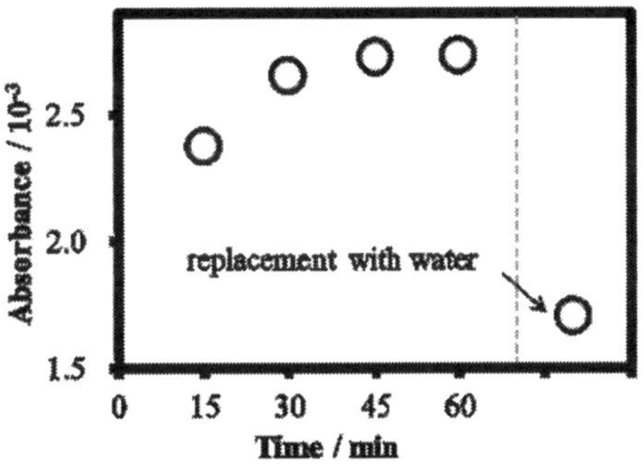

Fig. 6. v_aCH_2 absorbance vs. time on the DTAB solution and water.

Figure 5 shows the IER spectra of the d-STA/PAA/AuNP/DTAB film on the DTAB solution and water. The times shown in this figure represent the elapsed time after injecting the DTAB solution into the water underneath the d-STA/PAA/AuNP film. The peaks corresponding to the CD_2 stretching bands of d-STA and the CH_2 stretching bands of DTAB can be observed in the IER spectra. Figure 6 shows a plot of the intensity of the antisymmetric CH_2 stretching band of DTAB against the elapsed time on the DTAB solution and on water. Upon adding DTAB molecules to the water, the peak corresponding to DTAB appeared quickly and continued to gradually increase for several tens of minutes. This result indicates that DTAB was quickly adsorbed underneath the AuNPs, with saturation reached after several tens of minutes. The DTAB peak persisted even after replacing the DTAB solution with water, although the intensity decreased. This decrease is probably associated with the elimination of extra DTAB molecules that were not adsorbed on the AuNPs. However, the retention of this peak indicates that the DTAB molecules adsorbed tightly onto the AuNPs through ionic bonding between the quaternary ammonium ion and carboxylate ion groups of DTAB and AuNPs, respectively. The wavenumber of the antisymmetric CH_2 stretching band in the d-STA/PAA/AuNP/DTAB film at 2921 cm^{-1} was lower than that of DTAB molecules at the air–water interface (2924 cm^{-1}); this indicates that DTAB molecules are more ordered on AuNPs than on the water surface. On the other hand, following addition of DTAB, the intensity of the d-STA band decreased slightly and the wavenumber increased (2194 cm^{-1}), indicating that d-STA molecules on the AuNPs became somewhat disordered on adding DTAB. However, as the increase in the wavenumber is small and there is little difference between the TEM images before and after adding DTAB, we believe that Janus nanoparticles could be prepared properly using the present method.

We were able to prepare Janus particles with two kinds of ligands spatially separated at the air–water interface by means of a mild (using little organic solvent) procedure. The robustness of the polyion complex in the Langmuir monolayer likely enables this procedure. We could also estimate the structural arrangement of the ligands during and after the preparation of the monolayer. These spectral results indicate that this preparation method can be used to successfully fabricate Janus particles.

4 Conclusions

In this study, we prepared AuNP monolayers covered with DTAB and Langmuir monolayers of polyion complexes of d-STA and PAA at the air–water interface. The IER spectra of the monolayers reordered during and after preparation of the monolayer confirmed that the adsorption of both d-STA and DTAB molecules on the AuNPs is completed in several tens of minutes. Moreover, the two ligands are arranged in a highly ordered fashion on the particles. While further work using other ligands needs to be discussed, this preparation method is a useful mild procedure for fabricating Janus particles.

References

1. M. C. M. Daniel, D. Astruc, Chem. Rev. **104**, 293 (2004)
2. A. R. Tao, J. Huang, P. Yang, Acc. Chem. Res. **41**, 1662 (2008)
3. D. V. Talapin, J.-S. Lee, M. V. Kovalenko, E. V. Shevchenko, Chem. Rev. **110**, 389 (2010)
4. T. Kawai, D. J. Neivandt, P. B. Davies, J. Am. Chem. Soc. 122, 12031 (2000)
5. T. Hassenkam, K. Nørgaard, L. Iversen, C.J. Kiely, M. Brust, T. Bjørnholm, Adv. Mater. **14**, 1126 (2002)
6. K. M. Mayya, A. Gole, N. Jain, S. Phadtare, D. Langevin, M. Sastry, Langmuir. **19**, 9147 (2003)
7. K. M. Mayya, N. Jain, A. Gole, D. Langevin, M. Sastry, J. Colloid Interface Sci. **270**, 133 (2004)
8. S. Pang, O. Tetsuya, W. Tomoyuki, T. Kondo, T. Kawai, J. Colloid Interface Sci. **285**, 634 (2005)
9. L. -H. Chen, A. Dudek, Y. -L. Lee, C. -H. Chang, Langmuir. **23**, 3123 (2007)
10. A. Mogilevsky, R. Jelinek, Langmuir. **27**, 1260 (2011)
11. S. Tatur, A. Badia, Langmuir. **28**, 628 (2012)
12. K. D. Comeau, M. V. Meli, Langmuir. **28**, 377 (2012)
13. V. Sashuk, R. Hołyst, T. Wojciechowski, M. Fiałkowski, J. Colloid Interface Sci. **375**, 180 (2012)
14. D. I. Babenko, A. A. Ezhov, D. S. Turygin, V. K. Ivanov, V. V. Arslanov, M. A. Kalinina, Langmuir. **28**, 125 (2012)
15. A. A. Torrano, Â. S. Pereira, O. N. Oliveira, A. Barros-Timmons, Colloids Surfaces B Biointerfaces. **108**, 120 (2013)
16. B. P. Gagnon, M. V. Meli, Langmuir. **30**, 179 (2014)
17. S. Srivastava, D. Nykypanchuk, M. Fukuto, O. Gang, Nano Lett. **8**, 9857 (2014)
18. A. Walther, A. H. E. Müller, Chem. Rev. **113**, 5194 (2013)
19. G. Frens, Nat. Phys. Sci. **241**, 20 (1973)
20. M. Osawa, K. Ataka, K. Yoshii, Y. Nishikawa, Appl. Spectrosc. **47**, 1497 (1993)
21. E. Le Calvez, D. Blaudez, T. Buffeteau, B. Desbat, E, Langmuir. **17**, 670 (2001)

MATEC Web of Conferences **98**, 02001 (2017) DOI: 10.1051/matecconf/20179802001
ICOMF16-LB16

DNA multilayer film for loading and release of DNA oligomer

Y. Kamimura and *N.* Kato*

Graduate School of Science and Technology, Meiji University, Kawasaki 214-8571, Japan

Abstract. The Layer-by-Layer (LbL) method using polyelectrolytes is a typical strategy for preparing functional polymeric multi-layered films. Besides the electrostatic force between positively and negatively charged polyelectrolytes, the hybridization force between complementary DNA strands can be also used for building the DNA LbL films. Herein, we report the thermal responsible DNA LbL film that is designed to release specific DNA oligomers (10 bases) at an elevated temperature. Because the temperature and the ionic strength (IS) of the medium affect the hydrogen bonds in the double stranded DNA, the structural stability of the film against temperature and IS was investigated. The loading and release of the releasable oligomers were repeatedly performed on the DNA LbL film.

1 Introduction

The Layer-by-Layer (LbL) method was developed for the formation of thin polymer films by alternatively adsorbing positively and negatively charged poly-electrolytes on a charged surface [1]. With the aid of the template-assisted method, the LbL method can be applied for preparing functional polymeric capsules for smart drug carriers [2-4]. So far, the LbL method has been proven to be a powerful and versatile tool for various applications [5, 6].

Analogous to the use of positively and negatively charged materials as building blocks in the LbL films [1, 2], the DNA LbL films are also constructed from complementary pairs of oligonucleotides [7-14]. The physicochemical properties of the films and the hollow capsules made by the DNA LbL assembly have been investigated, and it was found that the properties can be controlled by the sequences of the DNA building blocks. Because the additional sequences not for forming the layered structure can be inserted in the building blocks, the functions such as the programmed degradation through competitive hybridization [14], the post-crosslinkable regions to increase the film stability [11], and the sequence to detect specific materials [15-17] can be easily introduced within the film.

The basic building blocks are the homopolymeric diblock nucleotides ($A_{15}G_{15}$ and $T_{15}C_{15}$, Figures 1(a, c)), and it had been pointed out that the slippage, where each DNA within the double helix can move along the homopolymeric region, assists the regular layer growth in the LbL assembly [7, 8]. This dynamic nature of the DNA stands might not only serve for the functionality such as the programed degradation [14] but also cause instability of the film structure [12].

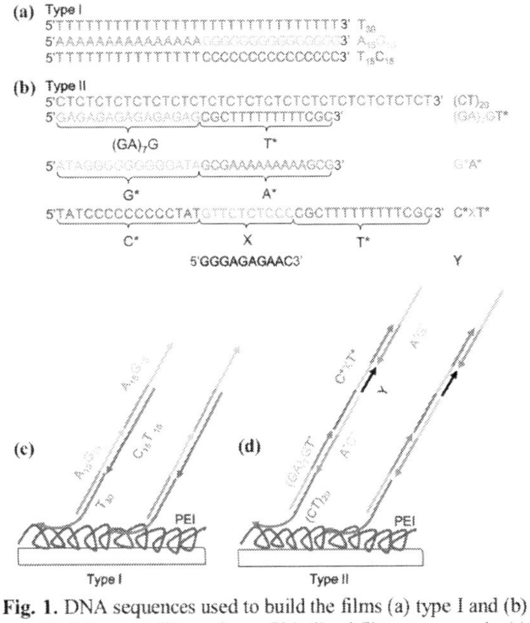

Fig. 1. DNA sequences used to build the films (a) type I and (b) type II. Schematic illustrations of idealized film structures in (c) type I and (d) type II. Polyethylenimine (PEI) is polycation that adsorbs T_{30} or $(CT)_{20}$ via electrostatic interaction.

Here, we report the DNA LbL film, whose structure was expected to be more stable than that built by the homopolymeric diblocks (Figures 1(a, c), type I). To supress the dynamic nature of the DNA strands, the homogeneity in the sequence was reduced, and the

* Corresponding author: nkato@meiji.ac.jp

proposed building blocks (type II) are shown in Figures 1 (b, d). Furthermore, a thermal responsive feature was introduced, i.e., the specific oligomer (oligo Y, Figures 1 (b)) can be released at an elevated temperature. The structural stability against temperatures and ionic strength was compared between the films built using the type I or the type II strands, and the loading and release of the releasable oligo Y were examined by means of quartz crystal microbalance (QCM). Gel electrophoresis was also performed to investigate whether the DNA building blocks form higher-order complexes in their solution.

2 Materials and methods

2.1 LbL assembly of DNA films

Pure water (> 18 MΩ cm) used in the present works was prepared in a Milli-Q system. The buffer solution (SSC buffer) for the DNA consisted of 50 mM sodium citrate with 500 mM NaCl and its pH was adjusted by NaOH to 6.5. The DNA strands were purchased from Hokkaido System Science Co., Ltd. and their purities were greater than 90%. Each DNA was dissolved in SSC buffer at a concentration of 4 μM and used for the assembly. The film growth and degradation were monitored using QCM. A quartz crystal oscillator with gold electrodes (9 MHz AT-cut) was used and its third overtone was monitored by a resonant frequency recorder (QCM934-000, Seiko EG&G Co., Ltd.). The solutions were injected in the recorder equipped with the oscillator at a flow rate of 1.26 μL/min and the temperature was kept at 25°C during the assembly. Because the large dissipation changes were observed during the DNA LbL assembly [8, 10], the Sauerbrey equation [18] does not hold. Therefore, the layer growth and degradation are qualitatively shown as a frequency change (Δf). To make the surface charge of the electrode positive, the aqueous solution of poly-ethylenimine (PEI, Mw ~ 25 000, 1mg/mL) was injected to the electrode for 10 min, followed by a rinse with SSC buffer for 20 min. On the PEI layer, the first DNA layer (T_{30} and $(TC)_{20}$ in type I and II, respectively) was adsorbed via electrostatic force. The DNA solution was injected for 3 min, allowing an incubation time of 10 min. This was followed by a rinse with SSC buffer for 7 min to remove any unhybridized strands. To assemble the film (type I), $A_{15}G_{15}$ and $T_{15}C_{15}$ were alternately deposited onto the T_{30} layer. In the case of type II, $(GA)_7GT^*$ (the second DNA layer) was deposited onto the $(TC)_{20}$ layer, followed by the alternate deposition of G*A* and C*XT*. This film is termed "type II without Y". After the formation of type II without Y, the oligo Y can be loaded, and such film is termed "type II with post-loaded Y". When G*A* and C*XT* hybridized with Y were alternately deposited onto the $(GA)_7GT^*$ layer, the resultant film is termed "type II with preloaded Y".

Because the oligo Y was designed to be released at an inflamed site, the films were exposed to an aqueous NaCl solution at a physiological ionic strength (IS) of 150 mM at 40°C that is higher than the normal body temperature. After the film formation in SSC buffer at 25°C, the 150 mM NaCl solution was injected for at least 25 min. The temperature was then increased to 40°C, allowing an incubation time for at least 15 min. After this incubation, the temperature was decreased to 25°C and the 150 mM NaCl solution was injected for 15 min, followed by an injection of SCC buffer for 15 min. The frequency of the QCM was then checked to compare with that after the film formation. This treatment with 150 mM NaCl at 40°C was performed not only for inducing the release of Y but also for examining the tolerance of the film during this treatment. To load the oligo Y in the film, the solution of Y was injected for 3 min at 25°C to the film, allowing an incubation time of 10 min. This was followed by a rinse with SSC buffer for 7 min to remove any unhybridized Y.

The temperature of the treatment with 150 mM NaCl was changed to 25, 30, or 35°C from 40°C to investigate the temperature dependence of the fraction of released oligo Y.

2.2 Gel electrophoresis

To detect the higher-order complexes of the strands, the native polyacrylamide gels (950 × 850 × 1 mm) were prepared using 40% (w/w) aqueous solution of acrylamide/N, N'-methylenebisacrylamide (19:1, Bio-Rad Laboratories Inc.). Ammonium persulfate and N, N, N', N'-tetramethylethylendiamine were used as a polymerization initiator and accelerator, respectively. Tris-borate buffer (78 mM, pH 8.3) with 877 μM EDTA disodium salt (TBE buffer) was used as an electrophoresis buffer. The DNA ladder (10 bp step, 1 μg/μL, Invitrogen™, Thermo Fisher Scientific Inc.) was denatured by heating at 70°C with formamide to create a single-stranded DNA ladder, and the final concentration of the ladder was 0.5 μg/μL. The sample DNA solutions were prepared using TBE buffer and the concentrations were from 0.2 to 2.0 μM. 10% (v/v) of glycerol was added to the ladder and the sample solutions before being applied to the gel. The pre-electrophoresis was made for 60 min at constant current of 20 mA. After applying the ladder and the sample solutions to the gel, the electrophoresis was made for 50 min at constant current of 80 mA. After the electrophoretic separation, the gel was stained by silver to visualize the separated bands using the staining kit (2D-silver stain-II, Cosmo Bio Co., Ltd.).

3 Results and discussion

3.1. Film assembly and stability

The changes in the frequency of the QCM during the film depositions using the type I strands or the type II stands without Y were observed. After the deposition of the first DNA layer, the frequency was set to 0, and the frequency changes (Δf) from the first DNA layer up to the ninth DNA layer are indicated in Figure 2. Type I showed a greater decrease in Δf after the film formation (Δf_{build}), indicating that the amount of the adsorbed DNA in type I is greater than type II. As pointed out in Ref. [12], in the case of type I, the multiple hybridization might occur due

DOI: 10.1051/matecconf/20179802001

to their homopolymeric nature, e.g., two $T_{15}C_{15}$ stands can hybridize to one $A_{15}G_{15}$ stand. However, such multiple hybridization was suppressed in type II due to the reduction in homogeneity in the sequences of the stands, resulting in the smaller decrease in Δf in type II.

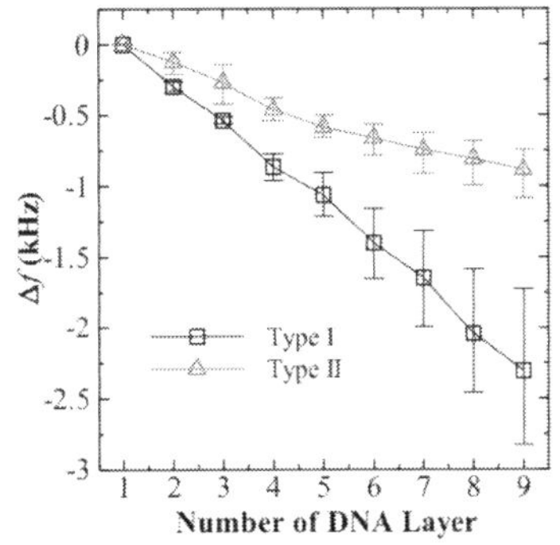

Fig. 2. Frequency changes during the film deposition using the type I strands (open square) and type II strands without Y (open triangle).

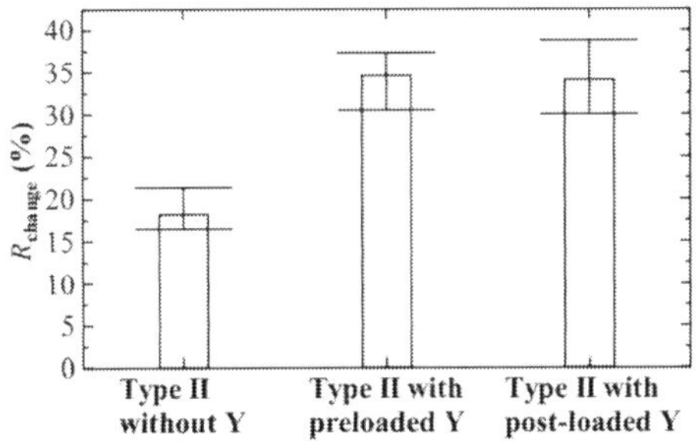

Fig. 3. Values of R_{change} observed in type II without Y, with preloaded Y, and with post-loaded Y. The films were composed up to the ninth DNA layer.

When the Δf after the treatment with 150 mM NaCl at 40°C (Δf_{treat}) was observed, a desorption of the DNA strands ($\Delta f_{build} < \Delta f_{treat}$) was observed in both types of films. The rate of change in the frequency after the treatment (R_{change}) was calculated as follows; $R_{change} = (\Delta f_{build} - \Delta f_{treat})/\Delta f_{build}$. The average values of R_{change} were 65.7 and 18.2% in type I and type II without Y, respectively. The lower R_{change} in type II indicates that the film stability was significantly improved by reducing the homogeneity in the sequences of the stands.

Although the stability was improved in type II, still the degradation in type II without Y (R_{change} = 18.2%) was significant enough to disturb the selective release of Y strands. In the idealized structure of type II built up to the ninth DNA layer with the oligo Y, the number ratio of the DNA strands in the film is $(GA)_7GT^*$:A*G*:C*XT*:Y = 1:4:3:3. From this ratio, the rate of mass decrease upon release of Y is calculated to be 14.2%. If the Δf is assumed to be proportional to the mass difference, the value of 14.2% corresponds to R_{change} upon release of Y. As shown in Figure 3, the values of R_{change} after the treatment with 150 mM NaCl at 40°C were around 34% in type II with post-loaded Y and type II with preloaded Y. These values were comparable to the sum of the value of R_{change} in type II without Y (18.2%) and the rate of mass decrease upon release of Y in the idealized film (14.2%). Thus, the film prepared using the type II strands was not stable enough to realize the selective release of Y triggered by the treatment with 150 mM NaCl at 40°C.

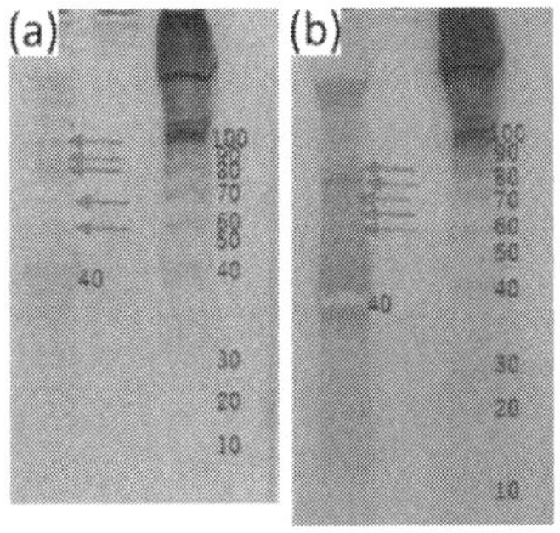

Fig. 4. (a) $(CT)_{20}$ at 0.2 μM in left lane and single stranded DNA ladder in right lane. (b) C*XT* at 2.0 μM in left lane and single stranded DNA ladder in right lane. To guide the eye, the arrows and the numbers of bases are indicated at the bands.

3.2 Higher-order complexes of the DNA strands

The DNA strands are able to form higher-order complexes, even if their sequences are not complementary each other [19]. We suspected that the higher-order complexes of the strands in the DNA solutions for the LbL assemble result in the lack of the structural stability of the film. To examine whether the strands shown in Figure 1(b) form the complexes, the electrophoresis using the native acrylamide gel was carried out. Together with the band at 40-mer, $(CT)_{20}$ and C*XT* exhibited the bands at larger values than 40-mer, as shown in Figure 4. In the case of $(GA)_7GT^*$ and G*A*, the evidence of the complexation was not observed. Because the DNA concentration and the IS of the DNA solution used in the electrophoresis (0.2 to 2 μM in 78 mM TBE buffer) were lower than those used in the film deposition (4 μM in 50 mM SSC buffer with 500 mM NaCl), the fraction of the complexes of $(CT)_{20}$ and C*XT* should be more significant during the film deposition. Therefore, the complexes of the strands should be embedded in the film. Such strands in the embedded complexes were dissociated from the film upon the treatment with 150 mM at 40°C. This is one of the reasons for the instability of the film made of the type II strands.

3.3 Loading and release of oligo Y

To realize the film with the function that the oligo Y was selectively released upon the treatment with 150 mM NaCl at 40°C, the film composed up to the ninth DNA layer with preloaded Y was exposed to the 150 mM NaCl solution at 40°C to destroy the complexes and remove not only the oligo Y but also the weakly hybridized stands. Into this "pre-treated" film, the oligo Y was reloaded, and the response to the treatment was investigated. Because the mass fraction of Y in the idealized film is 14.2% as mentioned above, the mass fraction of the film without Y must be 85.8%. If the mass change is assumed to be proportional to the Δf and the rate of the Δf after loading

Y is normalized to 100%, the rates of Δf after releasing Y and reloading Y are calculated to be 85.8 and 100%, respectively. When the cycle of reloading and release of Y is repeated, the rate of Δf should change as shown in Figure 5 (open triangle). The rate of Δf of the film after the pre-treatment was scaled to be 85.8%, and the change in the rate of Δf during two cycles of the reloading and release of Y was observed and the result is shown in Figure 5 (open square). The good agreement between the calculated and observed data during the two cycles indicates that the "pre-treated" film exhibited a good reproducibility in the reloading of Y and the selective release of Y upon the treatment with 150 mM NaCl at 40°C.

After reloading Y into the "pre-treated" film, the film was exposed to the 150 mM NaCl solution at different temperatures and the fractions of released Y were compared (Figure 6). As we designed, 100% of the loaded Y was released at 40°C. However, 80% was released at 35°C and 20% was released even at 25°C, indicated that the oligo Y is easily released under the normal body temperature. Thus, the current releasable oligo is not suitable enough for the selective release in the inflamed site and has to be improved its thermal stability by adjusting its sequence and the number of bases to increase the hybridization force.

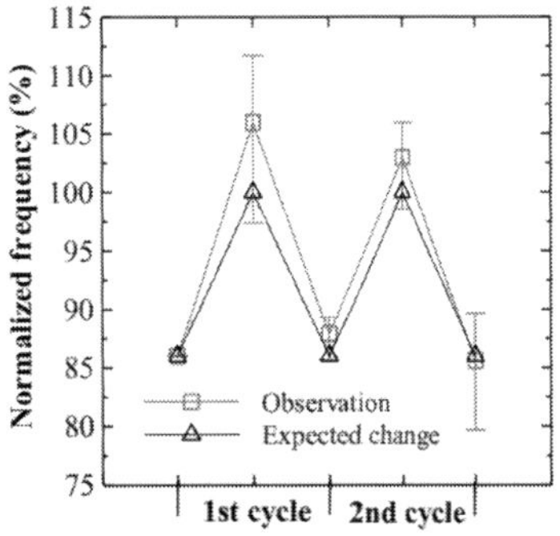

Fig. 5. Repetition of loading and release of Y after the pre-treatment. Open squares indicate the obverse data and open triangles indicate the calculated values.

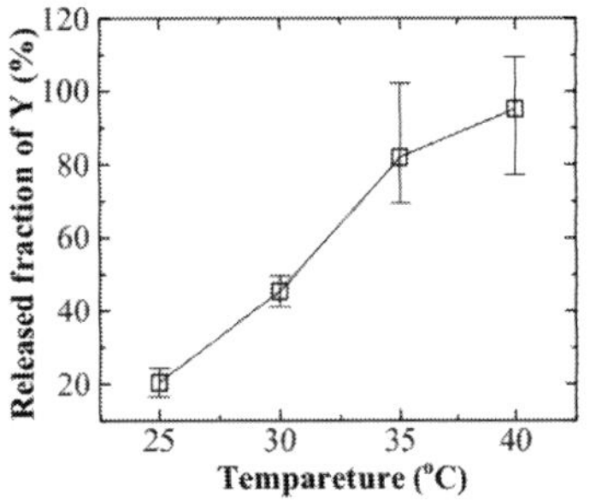

Fig. 6. Temperature dependence of the fraction of released Y. The film after reloading Y was exposed to 150 mM NaCl at different temperatures.

4 Conclusions

The film assembled by the proposed strands (type II, Figure 1(b)) exhibited a better structural stability toward the treatment with 150 mM NaCl at 40°C than the film assembled by the conventional homopolymeric dilbolck strands (type I, Figure 1(a)). However, the improvement in the stability was not enough for the selective release of the oligo Y. The lack of the film stability was ascribed to the higher-order complexes of the type II strands. After the elimination of these complexes from the film by the pre-treatment with 150 mM NaCl at 40°C, the loading and release of Y can be repeated as we expected. Although the oligo Y strands were partially released below the desired temperature, the basic concept of the thermal responsive DNA LbL film for the controlled release has been verified.

Acknowledgements

The authors acknowledge the financial supports from The Institute of Science and Technology, Meiji University and partially from JSPS KAKENHI Grant Number 16K01399. The authors also acknowledge significant experimental support from K. Hattori, N. Ueda, N. Kishi, Y. Inutsuka, and H. Shibuya.

References

1. G. Decher, Science **277**, 1232 (1997)
2. F. Caruso, R.A. Caruso, H. Mohwald, Science **282**, 1111 (1998)
3. S. De Koker, R. Hoogenboom, B.G. De Geest, Chem. Soc. Rev. **41**, 2867 (2012)
4. W. Tong, X. Song, C. Gao, Chem. Soc. Rev. **41**, 6103 (2012)
5. C. Picart, F. Caruso, J.-C. Voegel, *Layer-by-Layer Films for Biomedical Applications* (Wiley-VCH, Weihheim, Germany, 2015)
6. J.J. Richardson, M. Björnmalm, F. Caruso, Science **348**, 411 (2015)
7. A.P.R. Johnston, E.S. Read, F. Caruso, Nano Lett. **5**, 953 (2005)
8. A.P.R. Johnston, H. Mitomo, E.S. Read, F. Caruso, Langmuir **22**, 3251 (2006).
9. A.P.R. Johnston, F. Caruso, Angew. Chem. Int. Ed. **46**, 2677 (2007)
10. L. Lee, A.P.R. Johnston, F. Caruso, Biomacromol. **9**, 3070 (2008)
11. A.P.R. Johnston, F. Caruso, Small **4**, 612 (2008)
12. N. Kato, L. Lee, R. Chandrawati, A.P.R. Johnston, F. Caruso, J. Phys. Chem. C **113**, 21185 (2009)
13. L. Lee, A.P.R. Johnston, F. Caruso, Langmuir **28**, 12527 (2012)
14. L. Lee, A.P.R. Johnston, F. Caruso, Small **10**, 2902 (2014)
15. Y. Zheng, C. Yang, F. Yang, X. Yang, Anal. Chem. **86**, 3849 (2014)
16. H. Xing, N.Y. Wong, Y. Xiang, Y. Lu, Curr. Opin. Chem. Biol. **16**, 429 (2012)
17. M. Citartan, S.C.B. Gopinath, J. Tominaga, S.-C. Tan, T.-H. Tang, Biosens. Bioelectron. **34**, 1 (2012)
18. G. Sauerbrey, Z. Phys. **155**, 206 (1959)
19. E. Protozanova, R.B. Macgregor Jr., Biochem. **35**, 16638 (1996)

Phosphonate molecular layers on TiO_2 surfaces

P. Canepa[1], *I.* Solano[2], *S.* Uttiya[2], *G.* Gemme[3], *R.* Rolandi[1], *M.* Canepa[2], *O.* Cavalleri[1,*]

[1]Dipartimento di Fisica, Università di Genova, Via Dodecaneso 33, Genova, Italy
[2]Dipartimento di Fisica and OPTMATLAB, Università di Genova, Via Dodecaneso 33, Genova, Italy
[3]INFN, Sezione di Genova, Via Dodecaneso 33, Genova, Italy

Abstract. We investigated the deposition of $(HO)_2$-PO-S-$(CH_2)_{15}$-COOH on TiO_2. The phosphonate group has been chosen to serve as the anchoring moiety while the carboxylic group can be exploited for further coupling with antibacterial peptides. Native oxide-covered polished Ti samples were used as substrates for the molecular deposition. Previous to deposition, the substrates were plasma-cleaned in order to enrich their surfaces of hydroxyl groups, suitable for phosphonate coupling. The combined analysis of atomic force microscopy, X-ray photoemission spectroscopy and differential spectroscopic ellipsometry measurements indicates the formation of a submonolayer film made of irregular islands, one molecule thick.

1 Introduction

Titanium dioxide films are currently the object of intense investigation because of their wide range of applications, from photocatalysis, to photovoltaics and biomaterial development. In view of biomaterial development a crucial topic to be investigated is the interaction of the TiO_2 surface with organic and biological molecules. In particular, surface functionalization able to improve the antimicrobial and osteointegration properties of the TiO_2 surface is of large interest for implantology. Phosphonic acids [R-PO_3H_2] and their phosphonate ester derivatives [RPO_3R_2] are attractive anchoring groups for metal oxide surfaces [1-7].

In this work we investigated the deposition of a molecule bearing a phosphonate group at one end -the head- and a carboxylic group at the other end - the tail. We chose the carboxylic group as free tail since it can form amide ligation with the amine functions of biomolecules. In particular, we plan to couple antimicrobial peptides to the phosphonate layer in order to confer antibacterial properties to the functionalized TiO_2 surface.

2 Materials and Methods

Plates (10 mm x 10 mm x 1 mm) cut from a Ti sheet (99.6% purity, Advent Research Materials) were used as substrates. Before immersion in the phosphonate solution, the substrates were polished with SiC paper up to P4000 grit, then sonicated in acetone, ethanol and Milli-Q water.

Since surface bound-OH groups can serve as anchoring points for the formation of densely packed monolayers, some substrates were treated in an oxygen-plasma cleaner before immersion in the self-assembly solution in order to activate surface -OH groups. The properties of sonicated and plasma treated substrates were compared.

$(HO)_2$-PO-S-$(CH_2)_{15}$-COOH (ProChimia Surfaces, Poland), referred to as P-COOH in the following, was dissolved in 1,4-dioxane to a final concentration of 0.2 mg/ml. For molecular deposition, substrates were kept in the phosphonate solutions for 48 h at room temperature in the dark and subsequently rinsed in pure solvent.

Samples were characterized by Atomic Force Microscopy (AFM), X-ray Photoemission Spectroscopy (XPS) and Spectroscopic Ellipsometry (SE).

Tapping mode AFM measurements were performed using a Multimode/Nanoscope V system (Bruker) and Si cantilevers (OMCLAC160TS, Olympus).

The chemical analysis of the sample surfaces was carried out by XPS measurements using a 5600 Multi-Technique apparatus. An X-ray Al-monochromatized source ($h\nu$ = 1486.6 eV) was used. The binding energy scale was referenced by setting the C1s of adventitious carbon at 284.8 eV. The data have been deconvoluted after subtraction of a Shirley background.

SE measurements were performed using a rotating compensator ellipsometer (M-2000, J.A. Woollam Co. Inc.). The variations induced in the Δ spectra by the formation of an ultrathin layer are usually small but they can be highlighted using the difference spectra (for brevity δ-spectra) formalism. In practice differences between film-covered sample spectra and substrate ones

* Corresponding author: cavalleri@fisica.unige.it

($\delta\Delta = \Delta_{film} - \Delta_{substrate}$) taken in the same ambient are considered [8,9].

3 Results and Discussion

As a first step we investigated the effect of plasma cleaning treatment on the surface chemical composition of TiO_2 polished substrates. XPS analysis indicates that oxygen plasma produces several changes in the surface composition of the TiO_2 substrates. The decrease in C1s signal intensity and the increase in Ti2p signal intensity (data not shown) indicate that oxygen plasma removes some adventitious carbon. Moreover, plasma treatment modifies the O1s signal line-shape. Even plasma cleaning duration affects the surface chemical composition. Fig.1 shows that increasing the plasma cleaning time results in an increase of the O1s components in the region 531-533 eV. This spectral range includes the contribution of the hydroxyl group [10] which should covalently bind the organophosphorous molecules. We therefore decided to treat TiO_2 substrates in a plasma cleaner for 45 min before depositing P-COOH.

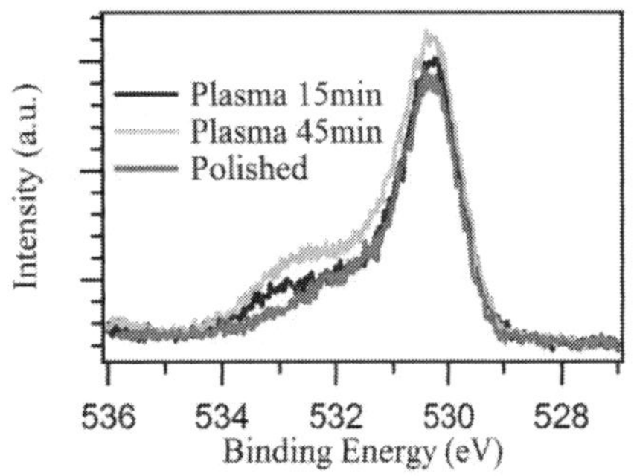

Fig. 1. XPS spectra of the O1s core level region of a polished TiO_2 surface as a function of the plasma cleaning time.

The analysis of AFM images indicates that oxygen plasma cleaning does not change the surface morphology significantly. At large scan size (several μm) polishing scratches can be observed (data not shown), but at smaller scales the surface morphology is relatively smooth and allows for monitoring morphological changes induced by molecular adsorption (see Fig.2).

AFM images taken after P-COOH deposition (Fig.2b) show significant changes compared to plasma cleaned substrates (Fig.2a). Irregular, partially coalesced islands form on the surface upon deposition. The average island height is around 1.5 nm, in agreement with the formation of monomolecular islands.

SE allows for a non-destructive fast check of the layer deposition. Fig.3a shows the Δ curves obtained before (blue line) and after (red line) P-COOH deposition. Preliminary simulations (not shown) suggest that the dip between 300-350 nm is related to the TiO_2 electronic gap and the position of the minimum depends also on the film thickness which can be estimated of the order of 10 nm. Simulations also indicate that the deposition of an ultrathin transparent layer on TiO_2 induces a decrease of the Δ values; in particular the decrease of the Δ values in the infrared region is roughly proportional to the thickness of the deposited film. Assuming a reasonable value for the refraction index of ~ 1.5, the decrease in Δ (see the δΔ spectrum in Fig.3b) fairly corresponds to the formation of a monomolecular layer. The tiny shift in the position of the Δ minimum upon deposition which produces the dip in the δΔ curve is also due to total film thickness increase upon P-COOH deposition. Such a dip must not be confused with a molecular absorption as observed, e.g., for cytochrome *c* monolayers on SiO_2 [11] or on gold substrates [9].

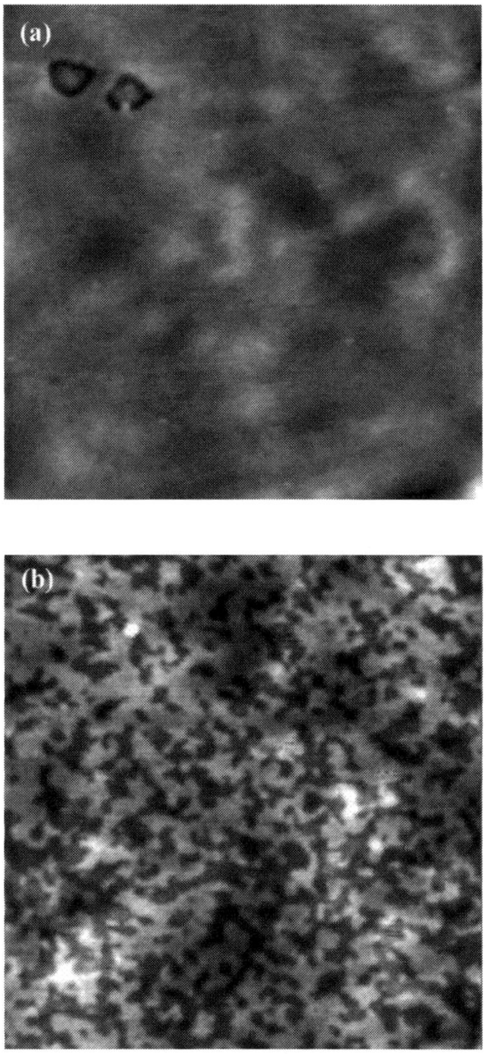

Fig. 2. Tapping mode AFM images of a plasma cleaned, polished TiO_2 surface, before (a) and after (b) P-COOH deposition. Scan size: 700 nm, z scale: 5 nm.

In order to gather more information about the nature of the islands, we employed XPS analysis to investigate the elemental composition of the surface upon P-COOH deposition. Fig.4 shows the XPS spectra of the (a) P2p, (b) S2p, (c) C1s and (d) O1s core level regions of a P-COOH layer deposited on plasma cleaned TiO_2.

The P2p signal is modelled with a single 2p doublet using two Voigt functions with a branching ratio of 1:2 and a spin orbit splitting of 0.86 eV. The $2p_{3/2}$ component is found at (133.6±0.2) eV as expected for the phosphonate group [10].

In a similar way, the S2p spectrum can be deconvoluted with a single doublet, with a branching ratio of 1:2 and a spin orbit splitting of 1.2 eV. The BE of the $2p_{3/2}$ component is found at (163.5±0.2) eV, in agreement with the S signal of physisorbed organosulphur compounds [12]. The presence of the S atom in the alkylic chain provides a further chemical signature of the molecular deposition.

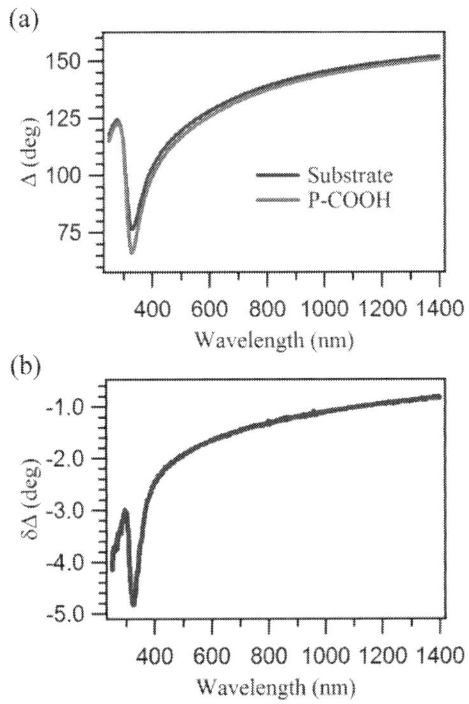

Fig. 3. (a) SE Δ spectra obtained on a TiO_2 plasma-cleaned substrate before (blue line) and after (red line) P-COOH deposition; (b) δΔ difference spectrum obtained as $\Delta_{after}-\Delta_{before}$

The C1s signal exhibits a main component with a shoulder on the high BE side and a second weak and broad signal at higher BEs. Three Voigt functions with the same FWHM have been fitted to the data. The main component, C1, at BE (284.8±0.2) eV is due to the alkylic chains and contains also the contribution of adventitious carbon. The second component, C2, at (286.5±0.2) eV can be attributed to C-S while the third component, C3, at (289.1±0.2) eV is related to the carboxylic group [3,10,13].

The O1s spectrum contains contributions from both the P-COOH molecule, the TiO_2 substrate and adventitious oxygen. The signal has a main component with a broad shoulder on the high BE side.

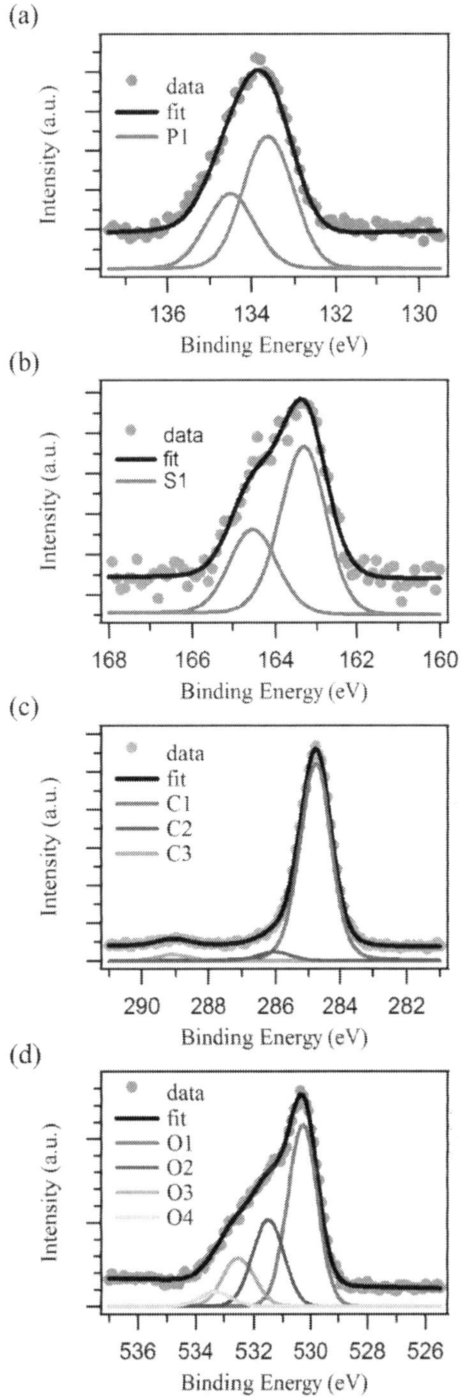

Fig. 4. XPS core level spectra of (a) P2p, (b) S2p, (c) O1s and (d) C1s regions of P-COOH layer deposited on plasma cleaned TiO_2 substrates

The data have been deconvoluted using four components with the same FWHM: O1 at (530.3±0.2) eV, O2 at (531.5±0.2) eV, O3 at (532.5±0.2) eV and O4 at

(533.3±0.2) eV. The O1 component can be assigned to the TiO_2 substrate [2,3,10,14]. Comparing with the expected BEs derived from literature, the O2 signal can be attributed to P-O-Ti and Ti-OH [10,14,15], the O3 component can be assigned to P=O [3,14,15] and C=O [3] while the O4 signal can be related to C-OH [3,14] and P-OH [3,15].

4 Conclusion

We investigated the adsorption on TiO_2 surfaces of an organophosphorous compound bearing a carboxylic end group (P-COOH) by combining AFM, SE and XPS. As concerns the substrate preparation, plasma cleaning was found to be useful to reduce adventitious carbon and to activate surface hydroxyl groups, suitable for anchoring phosphonates to the surface.

The AFM analysis indicates that P-COOH forms a sub-monolayer film on the TiO_2 surface with the formation of irregular islands on the surface. The analysis of the z profile of AFM images indicates the formation of islands of monomolecular height.

SE measurements show a decrease of the Δ values upon P-COOH deposition, further confirming the growth of the organic adlayer.

The analysis of the P2p, S2p, C1s and O1s XPS signals confirms the occurrence of molecular deposition as indicated by AFM and SE. However, at the present stage, we do not have conclusive evidence on the molecular binding mechanism and orientation on the surface. Indeed evidences have been reported in literature for the binding of carboxylic acid to TiO_2 surfaces [16,17]. On the other hand a previous study on the deposition of a ω-carboxyalkanephosphonic acid on TiO_2 reported, on the base of IR analysis, a preferential molecular binding through the phosphonate group [18]. Further analysis is therefore necessary to get a deeper insight in the molecule/surface binding mechanism. Ongoing angle-resolved XPS measurements will allow us to gain information on the molecular orientation within the layer, and therefore on the binding scheme.

A further aspect to be investigated will be the effect of post-deposition annealing on the film structure which has been reported to affect the binding as well [18].

Once clarified the anchoring mechanism, a step further will be the exploitation of unreacted carboxylic groups to anchor antimicrobial peptides through an amide bond in order to promote the antibacterial properties of the functionalized surface.

Acknowledgments

The authors thanks Prof. Sergio Thea for helpful discussions and for providing 1,4-dioxane and Dr. Sergio Moya for helpful discussions. The authors thank the group of Prof. Flavio Gatti for access to facilities for substrate preparation. The authors acknowledge funding from the University of Genova (PRA2014-FRA2015). S.U. acknowledges support from "Fondo Sociale Europeo Regione Liguria" (project n. DPU12UNIGE82/3600). P. Canepa and M. Canepa are not relatives.

References

1. S.P. Pujari, L. Scheres, A.T. M. Marcelis, and H. Zuilhof, Angew. Chem. Int. Ed. **53**, 6322 (2014)
2. S. Tosatti, R. Michel, M. Textor, N. D. Spencer, Langmuir **18**, 3537 (2002).
3. N. Adden, L. J. Gamble, D. G. Castner, A. Hoffmann, G. Gross, H. Menzel, Langmuir **22**, 8197 (2006)
4. A.Córdoba, M. Hierro-Oliva, M.A. Pacha-Olivenza, M.C. Fernández-Calderón, J. Perelló, B. Isern, M.L. González-Martín, M. Monjo, J.M. Ramis, ACS Appl. Mater. Interfaces **8**, 11326 (2016)
5. G. Mani, D.M. Johnson, D. Marton, V.L. Dougherty, M.D. Feldman, D. Patel, A.A. Ayon, C.M. Agrawal, Langmuir **24**, 6774 (2008)
6. N. Metoki, L. Liu, E. Beilis, N. Eliaz, D. Mandler, Langmuir **30**, 6791 (2014)
7. C. Viornery, Y. Chevolot, D. Léonard, B.-O. Aronsson, P. Péchy, H.J. Mathieu, P. Descouts, M. Grätzel, Langmuir **18**, 2582 (2002)
8. C. M. Herzinger, B. Johs, W. A. McGahan, J. A. Woollam, W. Paulson J. Appl. Phys. **83**, 3323 (1998)
9. C. Toccafondi, M. Prato, G. Maidecchi, A. Penco, F. Bisio, O. Cavalleri, M. Canepa, J Colloid Interf. Sci. **364**, 125132 (2011)
10. S. Bozzini, P. Petrini, M.C. Tanzi, S. Zürcher, S. Tosatti, Langmuir **26**, 6529 (2010)
11. C. Toccafondi, O. Cavalleri, F. Bisio, M. Canepa, Thin Solid Films **543**, 78 (2013)
12. G. Gonella, O. Cavalleri, S. Terreni, D.Cvetko, L. Floreano, A. Morgante, M. Canepa and R. Rolandi. Surf. Sci. **566-568**, 638 (2004)
13. O. Cavalleri, G. Gonella, S. Terreni, M. Vignolo, L. Floreano, A. Morgante, M. Canepa, R. Rolandi, Phys. Chem. Chem. Phys. **6**, 4042 (2004)
14. R. Hofer, M. Textor, N.D. Spencer, Langmuir **17**, 4014 (2001)
15. M. Gnauck, E. Jaehne, T. Blaettler, S. Tosatti, M. Textor, H.-J. P. Adler, Langmuir **23**, 377 (2007)
16. A.G. Thomas, K.L. Syres, Chem. Soc. Rev **41**, 4207 (2012)
17 C.L.Pang, R. Lindsay, G. Thornton, Chem. Rev. **113**, 3887 (2013)
18 E.S. Gawalt, M.J. Avaltroni, N. Koch, J. Schwartz, Langmuir **17**, 5736 (2001)

Graphene-based LbL deposited films: further study of electrical and gas sensing properties

A. Nabok[1,*], N. J. Walch[1], S. Dutton[1], F. Davis[1,2], S.P.J. Higson[2]

[1] Sheffield Hallam University, Materials and Engineering Research Institute, Sheffield, S11 7JN, UK
[2] Department of Engineering and Applied Design, University of Chichester, PO21 1HR, UK

Abstract. Graphene-surfactant composite materials obtained by the ultrasonic exfoliation of graphite powder in the presence of ionic surfactants (either CTAB or SDS) were utilised to construct thin films using layer-by-layer (LbL) electrostatic deposition technique. A series of graphene-based thin films were made by alternating layers of either graphene-SDS with polycations (PEI or PAH) or graphene-CTAB with polyanions (PSS). Also, graphene-phthalocyanine composite films were produced by alternating layers of graphene-CTAB with tetrasulfonated nickel phthalocyanine. Graphene-surfactant LbL films exhibited good electric conductivity (about 0.1 S/cm) of semiconductor type with a band gap of about 20 meV. Judging from UV-vis spectra measurements, graphene-phthalocyanine LbL films appeared to form joint π-electron system. Gas sensing testing of such composite films combining high conductivity of graphene with the gas sensing abilities of phthalocyanines showed substantial changes (up to 10%) in electrical conductivity upon exposure to electro-active gases such as HCl and NH_3.

1 Introduction

Graphene is one of the most popular materials of modern times with thousands of publications covering all possible aspects of the physical-chemical properties of graphene-based materials and their applications. Among several known technologies of graphene production, the method of ultrasonic exfoliation of graphite powder in the presence of surfactants [1, 2] is perhaps the least expensive. The resulting products, i.e. graphene-surfactant composite materials, have the useful properties of solubility in water and some organic solvents, thus enabling deposition by the simple thin film technologies of Langmuir-Blodgett (LB), spin-coating, and electrostatic layer-by-layer deposition [2, 3]. The electric properties of such films must be different from those of pristine graphene exhibiting high conductivity in the range of 10^6 S/cm caused by the large conjugated system of π-electrons with a zero band gap [4, 5]. Our previous study of LB and LbL graphene-surfactant composite films showed high electric conductivity and a temperature dependence of conductivity of semiconductor type (conductivity rising with temperature), though that required further verification.

The use of electrostatic LbL deposition offer another attractive possibility of alternating electrically charged graphene layers with other molecular layers, for example phthalocyanines. It is known that graphene itself is not ideal for gas sensing and requires functionalization with other molecular receptors to achieve the required gas sensitivity and selectivity [6, 7]. On the other hand, metal-phthalocyanines thin films are known by excellent gas sensitivity to electro-active gases though their conductivity is very poor [8]. A combination of highly conductive graphene with metal-phthalocyanines seems to be very attractive for development of novel gas sensors.

In the current research we continue studying electric conductivity of graphene-surfactant composite films with particular focus in temperature dependence of conductivity. The second part of this work is dedicated to making graphene-phthalocyanine LbL films and studying their optical and electrical properties including the effect of electro-active gases.

2 Experimental methodology

2.1 Samples preparation

Graphene composites were prepared by a multistage sonication process of graphite powder in the presence of ionic surfactants containing either negative or positive ionic groups, e.g. sodium dodecyl sulphate (SDS) and cetyl- trimethyl- ammonium bromide (CTAB), as described earlier [2, 3]. The resulted two graphene-surfactant composite materials, namely graphene-CTAB, and graphene-SDS (see schematic diagrams in figure 1a, b) appeared to be soluble in both water and some organic solvents, such as chloroform, and thus suitable for both

* Corresponding author: A.Nabok@shu.ac.uk

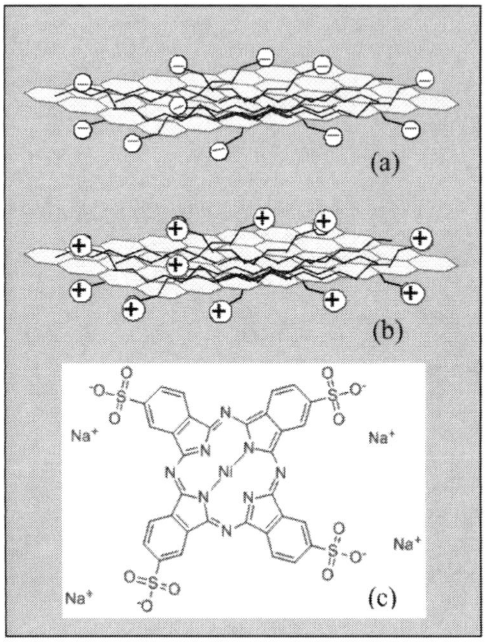

Fig. 1. Schematic representation of graphene-SDS (a) and graphene-CTAB (b). Chemical structure of NiPc derivative used (c).

layer-by-layer (LbL) electrostatic deposition and LB (Langmuir-Blodgett) deposition.

In this study, graphene-based films were deposited onto various substrates (silicon, glass, quartz, and gold interdigitated electrodes (from DropSens) using the electrostatic LbL process). Layers of graphene-CTAB were alternated with polyanions such as PSS, while graphene-SDS layers were alternated with polycations PEI. The surface of glass, quartz, and silicon after cleaning in pirana solution is usually hydrophilic and negatively charged due to the presence of OH^- groups, while gold electrodes were additionally treated overnight in 100 mM solution of mercapto-ethylene-sulfonate sodium salt in methanol to enhance the surface negative charge. Samples with the following layer sequence were produced: $(PEI/graphene\text{-}SDS)_n$, where n is the number of bilayers deposited, typically n= 1, 2, 3…

In the series of graphene-phthalocyanine samples graphene-CTAB was used in alternation with tetra-sodium sulphonate nickel phthalocyanine (NiPc-$(SO_3^- Na^+)_4$ or shortly NiPc) which chemical structure is shown in figure 1c. Electrostatic LbL films were deposited on the above mentioned substrates in the following sequences: $PEI/PSS/(graphene\text{-}CTAB/NiPc)_n$ or $PEI/(NiPc/graphene\text{-}CTAB)_n$, n =1, 2, 3... All chemicals used were purchased form Sigma-Aldrich.

2.2 Measurements techniques

The morphology of the LbL films deposited on pieces of Si wafer were studied using AFM (Nanoscope IIIa - Digital Technology/Bruker instrument) operating in tapping (scan-assist) mode using Bruker super-sharp tips with a radius of 2-4 nm, and the oscillation frequency of around 100kHz.

Optical absorption spectra of LbL films of graphene-phthalocyanine deposited on glass or quartz slides were recorded using Varian Carry 50 spectrophotometer. The thickness and dispersions of refractive index and extinction coefficient of the films produced were studied using J.A. Woollam M2000 spectroscopic ellipsometer.

Electrical DC measurements of graphene films deposited on interdigitated gold electrodes from DropSens (containing 100 fringes spaced by 5μm) were carried out using a Keithley 4200 electrometer in the temperature range from 77°K to 393°K maintained by Oxford Instrument cryostat Optistat DN2. Gas sensing tests of graphene-phthalocyanine samples deposited on interdigitated electrodes were performed using a 50 ml PTFE cell in which the saturated and 1:10 diluted vapours of NH_3 and HCl were injected with a 10 ml syringe. During such tests, time dependencies of current at fixed DC voltage of 1V were recorded using portable UNI-T UT61E multimeter.

3 Experimental results and discussion

3.1 Conductivity of graphene-surfactant composite LbL films

A typical AFM image of PEI/graphene-SDS film deposited silicon wafer is shown in figure 2. A single flake of about 500nm in size, which overlaps with other one or two smaller flakes, is clearly visible. Sectional analysis of this image allowed the estimation of the thickness of a single flake to be about 2 nm.

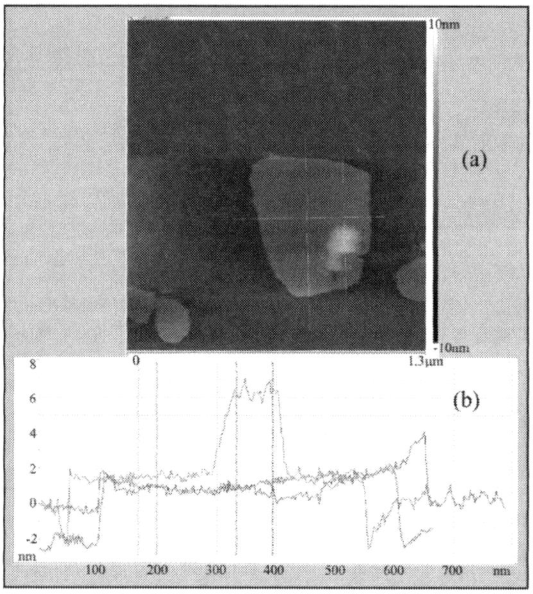

Fig. 2 AFM image of PEI/graphene-SDS layer on Si (a) and its section analysis (b)

Typical ellipsometry spectra of PEI/graphene-SDS film deposited on silicon are shown in figure 3. Deposition of layers of PEI and graphene-SDS result in substantial shift of Δ spectra, while Ψ spectra show little changes. This is quite common observation for thin transparent films, where Δ represent changes in the film thickness while Ψ reflects changes in the refractive index.

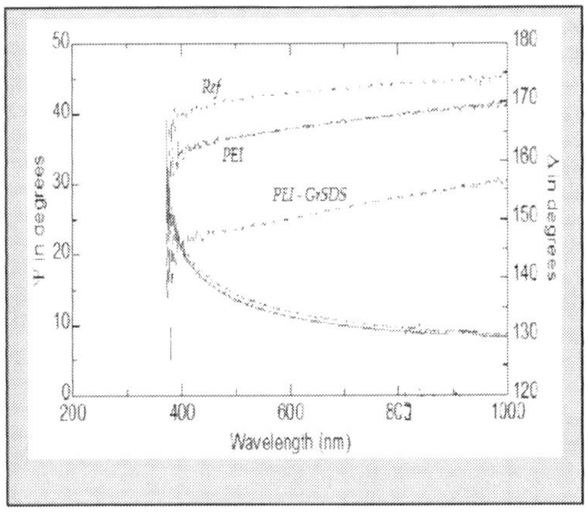

Fig. 3 Spectroscopic ellipsometry data for PEI/graphene-SDS layer on silicon.

Ellipsometry data fitting based on the Lorenz oscillator model for graphene layers [3] yields the thickness increment of 2.5 nm for PEI/graphene-SDS bilayer which is similar to what was observed directly with AFM.

All graphene-surfactant composite films studied before [3] and recently show Ohmic behaviour as illustrated in figure 4. The dependence of the current against the number of PEI/graphene-SDS bi-layers deposited is non-linear since the surface coverage in the first couple of layers is less then 100%. Then it stabilises and the conductivity of graphene-surfactant LbL films reaches 0.1 S/cm which is appreciable though it is not as high as that for pristine graphene films [4].

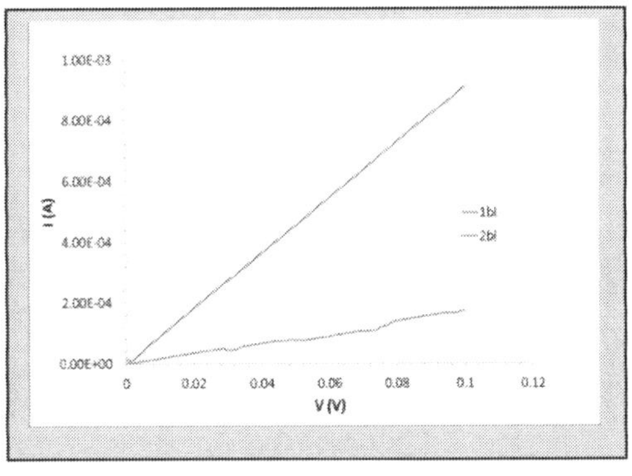

Fig. 4 Typical IV-characteristics of PEI/graphene-SDS films

Obviously, the presence of surfactants physically separates graphene flakes from each other and thus creates potential barriers for conductivity. Subsequently, the mechanism of conductivity must turn into the thermally activated type with conductivity values reduced. Such transformations are apparent form the temperature dependence of conductivity in figure 5.

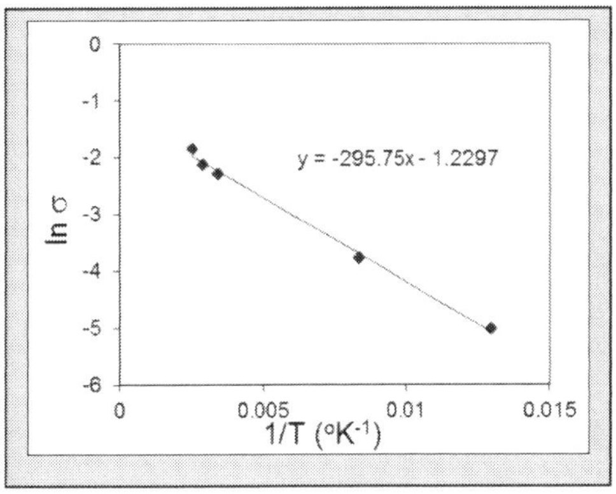

Fig. 5 Temperature dependence of conductivity of PEI/graphene-SDS films.

The conductivity is clearly of semiconductor type following a general formula $\sigma = \sigma_0 \exp(-E_a / kT)$. The activation energy E_a which could be interpreted as a semiconductor band-gap E_g was found to be of 20 meV. This is a very small energy comparable with kT at room temperature but it is not zero as compared to pristine graphene [4].

3.2 Gas sensing properties of graphene-nickel phthalocyanine LbL films

Composite LbL films of graphene-CTAB alternated with NiPc were studied with UV-vis absorption spectroscopy, and the results obtained are shown in figure 6. The main spectral features are typical of NiPc with characteristic spectral absorption bands below 400nm known as the Soret band and the Q-band between 550 and 750nm; the contribution of graphene to the spectra is almost uniform over the whole spectral range and resulted in raising the baseline. However, the presence of graphene has resulted in intriguing transformations of NiPc Q-band. The layers of NiPc deposited either on graphene-CTAB (fig. 6a) or on PEI (fig. 6b) exhibit a Q-band consisting of Qx (615nm) and Qy (670nm) sub-bands with the Qx peak higher that Qy [8].

(a)

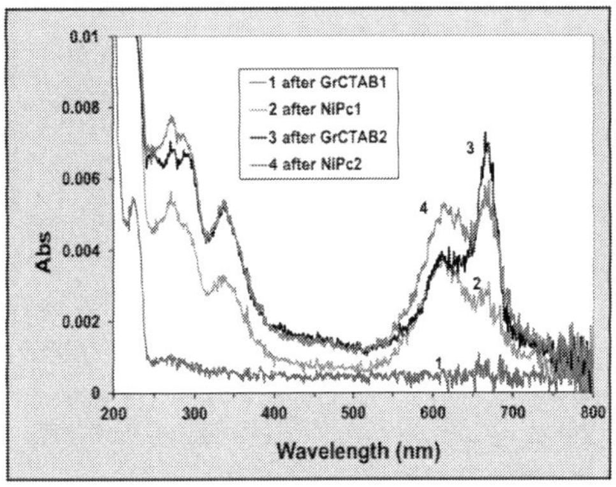

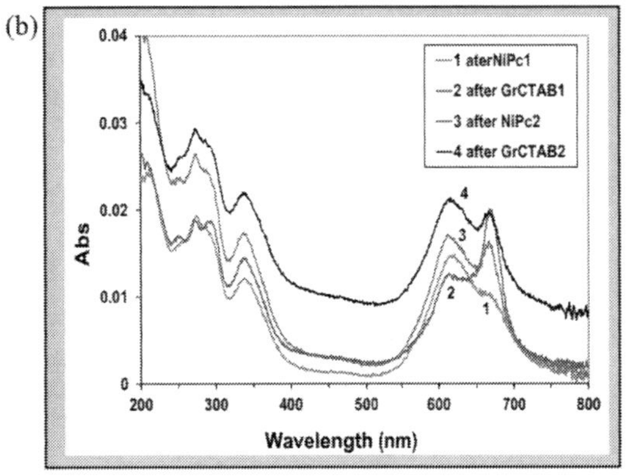

Fig. 6 UV-vis absorption spectra of LbL films with the layer sequence of graphene-CTAB/NiPc (a) and NiPc/graphene-CTAB (b)

This is similar to the NiPc spectrum in aqueous solution and typical for insulated NiPc molecules. Deposition of graphene-CTAB on top of NiPc causes a reversal of amplitudes of Qx and Qy with Qy becoming more intense somehow similar to the spectra of metal free phthalocyanines [8]. The observed spectral changes are the indication of interaction of π-electrons of NiPc molecules with those of graphene, most-likely resulting in the ionisation of Ni and subsequent formation of metal-free phthalocyanine molecules [8, 9].

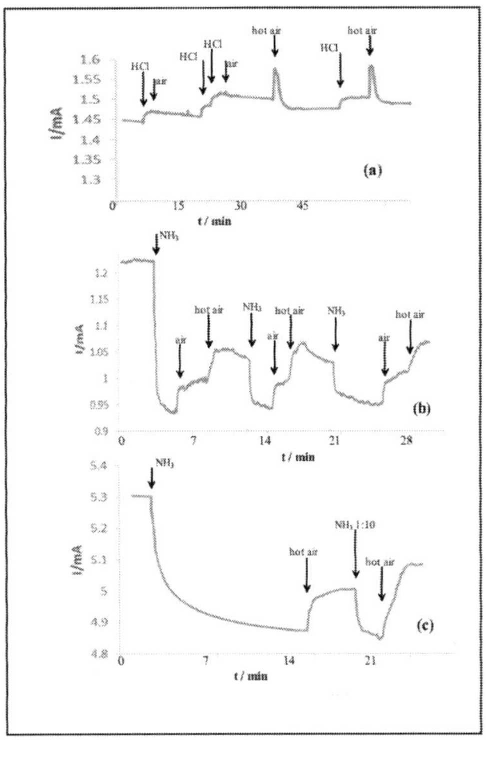

Fig. 7 The responses of conductivity of graphene-CTAB/NiPc films to injection of gases: HCl (a) and NH_3 (b, c). Injections of respective vapours, air, and hot air are shown with arrows.

Such a situation (when π-electrons of phthalocyanine and graphene overlap) is highly desirable for improving gas sensing properties of these composite materials. Indeed, electrical measurements of NiPc/graphene-CTAB films confirmed this.

I-V characteristics of graphene-CTAB/NiPc LbL films are very similar to those without NiPc. The current recorded at 1V DC bias is modulated by injection of vapours of HCl and NH_3 as shown in figure 7. Notably, HCl saturated vapours caused the conductivity to increase by about 3.5% (fig. 7a) while saturated NH_3 vapours caused its decrease initially by 30% though following exposures resulted in about 15% decrease (fig. 7b). The response is fast (just few seconds) but the recovery is much slower and not complete. Additional heating (using hot air) was required to improve the sensor recovery. The sensor response is reproducible (fig. 7b) and dependant on the vapour concentration; 1:10 dilution of saturated NH_3 led to much smaller sensor response (fig. 7c).

4 Conclusions

The study of electrical characteristics of graphene-surfactant composite LbL films confirmed the thermally activated mode of conductivity typical for semiconductors with the band-gap being about 25meV. Thin LbL films produced by alternation of positively charged layers of graphene-CTAB with negatively charged tetra- sodium-sulphonate NiPc layers demonstrated the interaction of π-electrons of NiPc moiety with those of graphene. As a result, the gas sensing properties of such composite materials were improved by combining the high conductivity of graphene with the gas sensing functionality of NiPc. A substantial response of conductivity of graphene thin films functionalized with NiPc to HCl and NH_3 vapours was observed. Further work is currently underway to study gas sensing properties of graphene-phthalocyanine (or porphyrin) composite films in more detail.

References

1. S.M. Notley, Langmuir, **28**, 4 (2012)
2. N.J. Walch, F. Davis, N. Langford,, J.L. Holmes, S.D. Collyer, S.P.J. Higson, Anal. Chem. **87**, 18 (2015)
3. N. J. Walch, A. Nabok, F. Davis, S. P. J. Higson, Beilstein J. Nanotechnol. , **7** (2016)
4. K.S. Novoselov, A.K. Gaim, S.V. Morozov, D. Jiang, Y. Zhang, S.V. Dubonos, I.V. Grigorieva, A.A. Fisov, Science, **306** (2004)
5. E.W. Hill, A. Vijayaragahvan, K. Novoselov, IEEE Sens. J., **11** (2011).
6. S. Basu, P. Bhattacharyya, Sens. Act. B Chem., **173** (2012)
7. P.Leenaerts, B. Partoens, F.M. Peeter, Phys. Rev B, **77** (2008)
8. J. Simon, J.-J. Andre, *Molecular Semiconductors*, Springler-Verlag, Berlin Heidelberg, 1985
9. D.W. Clack, J.R Yangle, Inorganic Chemistry, **11**, 8 (1972)

Heteroepitaxial growth of SiC films by carbonization of polyimide Langmuir-Blodgett films on Si

S.I. Goloudina[1,*], V.V. Luchinin[1], V.M. Pasyuta[1], M.F. Panov[1], A.N. Smirnov[2], D.A. Kirilenko[2], T.F. Semenova[3], V.P. Sklizkova[4], I.V. Gofman[4], V.M. Svetlychnyi[4] and V.V. Kudryavtsev[4]

[1] St. Petersburg State Electrotechnical University, 197376, 5 Prof. Popov st., St. Petersburg, Russia
[2] Ioffe Institute, RAS, 194021, 26 Politekhnicheskaya, St Petersburg, Russia
[3] St. Petersburg State University, 199034, 7-9 Universitetskaya nab., St. Petersburg, Russia
[4] Institute of Macromolecular Compounds, 199004, 31 RAS, Bol'shoy pr., St.Petersburg, Russia

Abstract. High quality single crystal SiC films were prepared by carbonization of polyimide Langmuir-Blodgett films on Si substrate. The films formed after annealing of the polyimide films at 1000°C, 1100°C, 1200°C were studied by Fourier transform-infrared (FTIR) spectroscopy, X-ray diffraction (XRD), Raman spectroscopy, transmission electon microscopy (TEM), transmission electron diffraction (TED), and scanning electron microscopy (SEM). XRD study and HRTEM cross-section revealed that the crystalline SiC film begins to grow on Si (111) substrate at 1000°C. According to the HRTEM cross-section image five planes in 3C-SiC (111) film are aligned with four Si(111) planes at the SiC/Si interface. It was shown the SiC films (35 nm) grown on Si(111) at 1200°C have mainly cubic 3C-SiC structure with a little presence of hexagonal polytypes. Only 3C-SiC films (30 nm) were formed on Si (100) substrate at the same temperature. It was shown the SiC films (30-35 nm) are able to cover the voids in Si substrate with size up to 10 μm.

Introduction

Among various semiconductor materials, silicon carbide (SiC) is an attractive material for high power, high frequency and high temperature microelectronics, owing in part to its wide band gap, high thermal conductivity, high break down field and high saturation velocity. Compared with Si or GaAs, it is chemically inert, extremely hard, radiation-resistant, and highly wear-resistant making it possible to fabricate microsensors and microactuators for harsh environments such as high temperature, aggressive media, and radiation-exposed environments. Additionally, SiC is inert to most chemicals at room temperature and has been found to be biocompatible with inert to bacterial growth, transparent to visible light, and show UV wavelength absorption. Furthermore, owing to its large ratio of Young's modulus to density, SiC has also attractive interest for use in ultra-high-frequency nanoelectromechanical system (NEMS) for wire-less signal-processing systems. However, this material has not been widely used because of the difficulties in growing high quality crystals and etching the material to form required pattern.

The motivation for the heteroepitaxial growth of SiC-on-Si has been to provide relatively inexpensive and large-area substrates of SiC for electronic devices and microelectromechanical system. Due to its large energy band and other promising semiconductor properties, 3C-SiC has potential for high performance as a wide band-gap emitter to the SiC/Si heterojunction bipolar transistor. Also SiC film on Si can be used as a buffer layer for the subsequent heteroepitaxial growth of gallium nitride and other group III-nitrides, which has application in blue and violet light-emitting diodes and lasers. However, the growth of SiC films on Si is one of the most difficult challenges of heteroepitaxy due to the large mismatches in lattice constant (20%) between SiC and Si, and the difference in the thermal expansion coefficient of SiC (4.6×10^{-6}/°C) and Si (4.2×10^{-6}/°C). The usual techniques to grow SiC-on-Si are chemical vapor deposition (CVD), laser sputtering, molecular beam epitaxy, and carbonization of a polyimide Langmuir-Blodgett (LB) film on silicon substrate. Currently, the high quality crystalline SiC films on Si can be fabricated by CVD method [1, 2], thickness of such films is often more than several microns. If the films thickness is less than one

* Corresponding author: goloudina@mail.ru

micron, it usually contains a large number of defects [3, 4].

The aim of this work was to prepare SiC films on Si by carbonization of rigid-rod polyimide (BPDA-oTD) LB films and studying their structure and morphology.

Materials and Methods

As polyimide prepolymer, we used polyamic acid alkylammonium salt (PAAS) synthesized on the basis of 3,3',4,4'-diphenyltetracarboxylic acid dianhydride and o-tolidine with tert-amine o,o',o''-trihexadecanoyltriethanolamine. The PAAS forms well-ordered and stable monolayers at the air-water interface. The PAAS monolayers were deposited on the silicon support at the surface pressure 30 mN/m. Y-type of deposition was observed. The transfer rate was 0.5 cm/min. [5].

PAAS LB films of 81, 121, 141 layers were deposited on Si(111) and Si(100) substrates. The polyimide (PI) films were then formed by thermal imidization of PAAS LB films (figure 1). The PI films had thickness of 47, 54 and 86 nm, respectively.

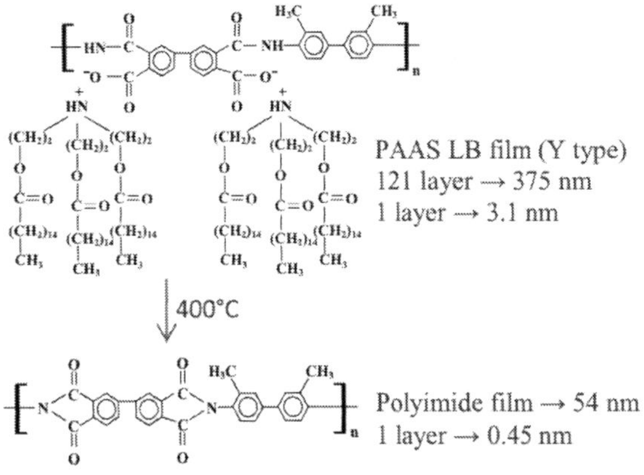

Fig. 1. The scheme of obtaining polyimide film.

For forming SiC the PI films on Si were annealed by two steps. Firstly the samples were heated to the temperature of 1000°C with the velocity of 10 degrees/min. and kept at this temperature during 1 h. After cooling down the samples were studied and then the films were heated again by the quick thermal heat annealing during 3 min. at the temperatures of 1100°C or 1200°C. The annealing was carried out in a vacuum (10^{-5}mm. of m.c.). The films formed after annealing were studied by Fourier transform-infrared (FTIR) spectroscopy, X-ray diffraction (XRD), Raman spectroscopy, high-resolution transmission electron microscopy (HRTEM), transmission electron diffraction (TED), and scanning electron microscopy (SEM).

The transmission spectrums of SiC films were obtained by using a Nicolet 6700 FT-IR Spectrometer. Raman spectra were acquired by using a Horiba Jobin Yvon T64000 micro-Raman system with a confocal microscope. The samples were excited by a continuous wave diode pumped Nd:YAG (λ_{ex}=532 nm) laser. Raman spectra were measured at room temperature, with the spot size on the sample of about 1 µm. The excitation power was limited to 1 mW to avoid sample heating. A standard Si(111) wafer was used for calibration of the micro-Raman system. X-ray investigations were carried out on a diffractometer with a Rigaku Ultima IV (CuKα radiation). TEM studies were performed by using a Jeol JEM-2100F transmission electron microscope (accelerating voltage 200 kV, point-to-point resolution 0.19 nm) equipped with an energy-dispersive X-ray spectrometer Oxford Instruments INCA. High resolution images were obtained in conventional bright-field mode using objective aperture of the optimal size. Specimens for TEM were prepared by mechanical polishing with subsequent ion milling using Ar^+ at 4 kV. The surface morphology of SiC was examined by FEI Quanta Inspect SEM.

Results and Discussion

The infrared spectrum of the films after annealing at 1000°C is shown in figure 2. This spectrum was obtained at a "reflection" regime. Only one absorption peak is seen at 796 cm^{-1}, which indicate the TO phonon of the 3C-SiC phase.

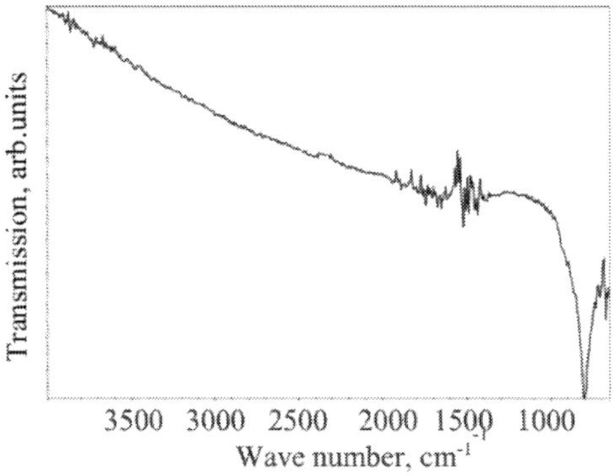

Fig. 2. FTIR spectrum of SiC films

The interesting details of SiC formation on Si were found by Raman spectroscopy. The results show that after heat treatment at 1000°C the films on Si consist mainly of amorphous carbon (figure 3 a). After annealing at 1100°C the films contain both the carbon and SiC phases (figure 3b). Mainly the SiC phase was formed at 1200°C (figure 3 c, d).

At the same time, the cross section TEM image of the film formed at 1000°C shows the presence of SiC nanocrystals on the Si surface (figure 4). Furthermore, it is clearly seen that five planes in 3C-SiC(111) structure are aligned with four Si (111) planes at the SiC/Si interface.

Formation of high quality SiC crystalline phase at 1000°C was confirmed by X-ray diffraction (figure 5). The peak 1 on the XRD spectrum corresponds to 3C-SiC(111) structure or H-SiC (table 1). The FWHM of this peak is 0.060° which indicates the single crystal SiC phase formation on Si [1].

DOI: 10.1051/matecconf/20179804002

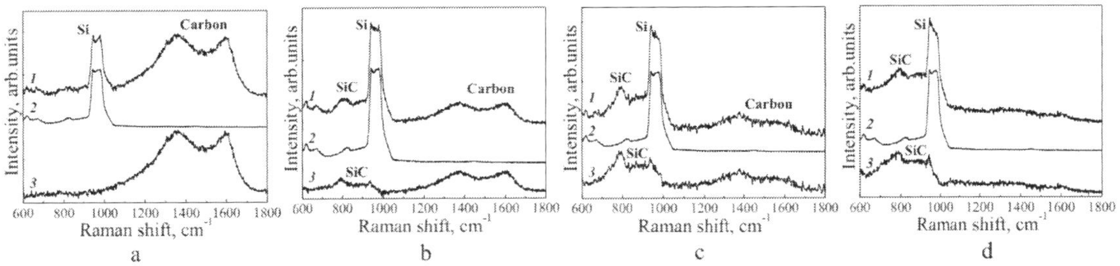

Fig. 3. Raman spectra of films formed on Si substrate at various temperature: Si(111) - 1000°C (a), 1100°C (b), 1200°C (c); Si(100) - 1200°C (d): *1* –SiC/Si, *2* – Si, *3* – difference spectrum (the spectrum *1* – the spectrum *2*).

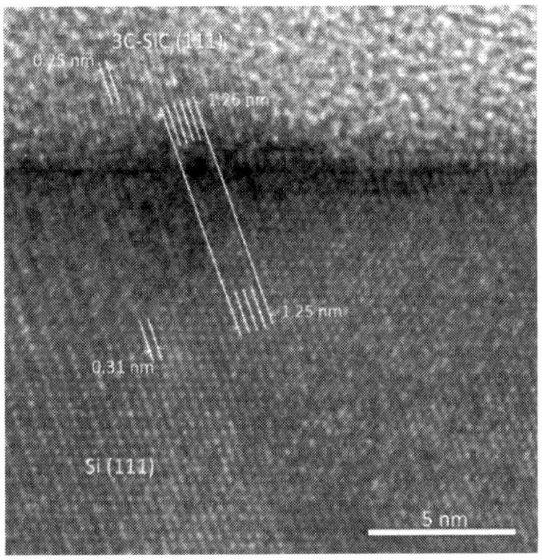

Fig. 4. Cross section TEM image of a SiC nuclei on a Si surface.

Figure 6 shows cross section images and TED of SiC films formed on Si at 1200°C. For preparing these films, PI films containing 121 layers were used. As can be seen in figure 6, the thickness of the SiC film is 35 nm on Si(111), and 30 nm on Si(100).

The SiC film grown on Si(111) has a 3C-SiC structure with a small presence of a hexagonal phase. Only 3C-SiC films were formed on Si(100) substrate. It was shown that the SiC film on Si(100) consist of two layers. The lower layer (10 nm) has crystalline structure and the upper layer (20 nm) is polycrystalline.

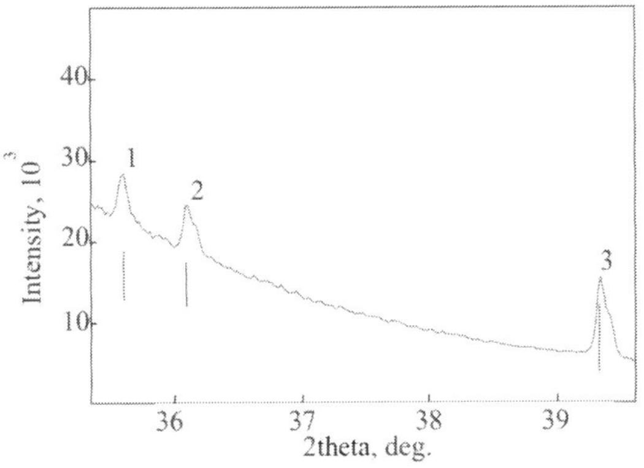

Fig. 5. XRD spectrum of the films after annealing at 1000°C.

Table 1. The characteristics of XRD spectrum.

N	2theta (deg.)	d (ang.)	FWHM (deg.)	Phase name
1	35.619	2.518	0.060	3C-SiC or H-SiC
2	36.100	2.486	0.080	8H-SiC
3	39.308	2.290	0.080	unknown

The surface morphology of SiC films was examined by SEM. Figure 7 (a, b) shows the presence of faced voids at the SiC/Si interface. Most of the voids are closed by SiC film. It was observed that the shape of voids depends on the symmetry of the substrate surface. The facets of voids are parallel to the lowest surface energy faces, i.e. {111} planes (figure 8). Thus, it is approximately square in shape on Si (100) and triangular on Si (111) substrates. The formation of the voids at the SiC/Si interface is caused by the diffusion of the Si atoms from the Si substrate.

Conclusion

In the present work we demonstrate an easy strategy to produce high quality single crystal SiC films on Si through carbonization of polyimide Langmuir-Blodgett films. To the best of our knowledge, this is the first report of the growth of single crystal SiC hetero-epitaxial film on Si with thickness of about 30 nm. The SiC {111} lattice planes are well aligned with those of the Si substrate. Moreover, every fourth Si plane is aligned with every fifth plane in 3C-SiC(111) structure at the SiC/Si interface. It was shown that the SiC films (30-35 nm), obtained at 1100°C and 1200°C, are able to close the voids in Si substrate. The size of these voids can reach 10 μm.

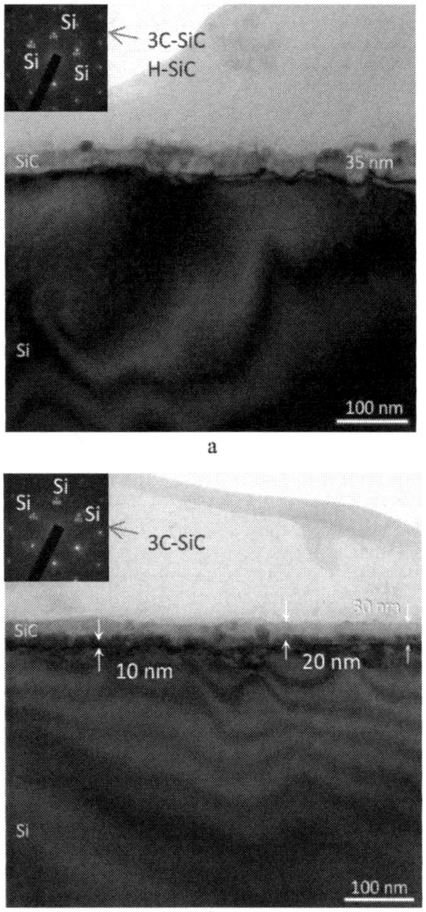

Fig. 6. Cross section images and TED of SiC films formed on Si at 1200°C: a) The SiC film (35 nm) on Si (111) substrate has a 3C-SiC structure with a small presence of a hexagonal phase; b) The SiC film on Si(100) substrate consist of two layers, the lower layer (10 nm) has 3C-SiC structure, and the upper layer (20 nm) is polycrystalline.

Fig. 7. SEM images of the SiC films covering the voids in Si (111) (a) and Si(100) substrates (b). The SiC films on the surface of Si substrate and the SiC films covering the voids have the different contrast. The broken SiC films are also observed.

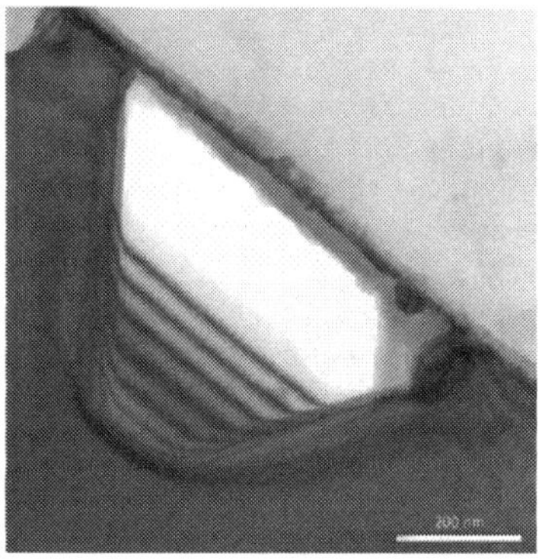

Fig. 8. Cross section TEM of SiC film with thickness 30nm, covering a pyramidal void in Si(100) substrate.

References

1. S. Madapura, A. J. Steckl, M. Loboda. Journal of Electrochemical Society. P.1197-1202. **146**(3) (1999)
2. C. Locke, R. Anzalone, A. Severino, C. Bongioro, G. Litrico, F. La Via, S.E. Saddow. *Materials Science Forum*. P.145-148. **615-617** (2009)
3. G.-S. Chung, K.-S. Kim. Bull. Korean.Chem.Soc. P.533-537. **28**(4) (2007)
4. A. Severino, R.Anzalone, M.Camarda, N. Piluso, F. La Via. *Materials Science Forum*. V.**711**, P.27-30 (2012)
5. S. I. Goloudina, V. V. Luchinin, V. M. Pasyuta, V.V Klechkovskaya, Yu. G. Baklagina, V.V. Rozanov, M.F. Panov, V.P. Sklizkova, V.V. Kudryavtsev. Crystallography Reports. P 468-476. **54**(3) (2009)

Mesoporous silica hollow capsules embedded with magnetic nanoparticles

S. Yoshikawa, *N.* Kato*, and *G.* Obara

Graduate School of Science and Technology, Meiji University, Kawasaki 214-8571, Japan

Abstract. Our method for synthesizing the mesoporous silica hollow capsules (MSHCs) was combined with the Layer-by-Layer method to integrate the magnetic nanoparticles (mNPs) onto the core template for the MSHC synthesis. The resultant magnetic MSHCs (mMSHCs) were monodisperse as the conventional MSHCs, and the mNPs were embedded on the inner surface of the mesoporous silica shell. The same mesoporous structure as in the MSHCs was confirmed in the mMSHCs and the properties of the mNPs were maintained even after being incorporated into the mMSHCs.

1 Introduction

The mesoporous silica hollow capsules (MSHCs) have great potential as drug carriers in drug delivery system, carriers of enzyme or catalysis, and adsorption and separation materials. To produce the MSHCs, numbers of methods based on the templated-assisted strategy have been proposed [1, 2]. We had also reported the methods for synthesizing the monodisperse MSHCs with uniform shell thickness [3, 4], and the loading and release of dyes had been demonstrated [3, 5]. In our methods, the surfactants and polystyrene (PS) nanoparticle were used as the templates to form the mesoporous structure in the shell and the hollow cavity, respectively.

In addition to the inherent feature of MSHCs, i.e., the large hollow cavity and the mesoporous network in the shell, further functionalized MSHCs such as luminescent and/or magnetic MSHCs (mMSHCs) are of interest for theranostic purposes [1, 6]. Specifically, the magnetic properties allow the magnetically-guided drug delivery for the therapy and the contrast-enhanced magnetic resonance imaging for the diagnosis. So far, various types of mMSHCs have been reported; the mMSHC incorporating the mNPs on the outer surface of the shell [7], inside the shell [8, 9], or in the hollow cavity [10-12] and the rattle-type mMSHC that encapsulates one magnetic particle in its hollow cavity [13-18].

Here, we aim to extend our method to produce the mMSHC that has the same uniformities in size and shape and the same mesoporous structure compared to the MSHCs reported before [3]. To incorporate the mNPs, the Layer-by-Layer (LbL) method [19] that had provided the magnetic hollow microspheres [20, 21] has been combined with our method, i.e., the multilayers of polyelectrolytes and mNPs were formed on the PS particle via the LbL method, followed by the sol-gel reaction to form the layer of silica/surfactant complex on the LbL layer. The coated particles were then calcined to pyrolyze the PS and surfactants, resulting in the mMSHCs.

2 Materials and methods

The mNPs were prepared based on the chemical co-precipitation mothed reported in the literature [22]. 100 mL of solution containing 17.4 mM $FeCl_3$ and 8.4 mM $FeCl_2$ was made using an aqueous HCl (0.2 M) as a solvent, followed by the addition of 5.4 mL of an aqueous NaOH (10 M). The solution was stirred at R.T. for 1 h and at 90 °C for another 1 h. With the aid of a magnetic separation, the mNPs were collected and redispersed in 6 mL of an aqueous sulfuric acid (9.5% w/w), followed by the a stirring at R.T. for 5 min to reduce the size of mNPs. By the magnetic separation, the mNPs were washed with water until the pH became neutral and redispersed in 100 mL of aqueous citric acid (0.3 M), followed by a stirring at 90 °C for 30 min. The dispersion medium was then exchanged to ethanol, and the sample was dried. All the processes were carried out under N_2 atmosphere.

The mNPs, polyethylenimine (PEI, Mw ~ 25 000), and poly(diallyl dimethyl ammonium chloride) (PDDA, Mw > 100 000) were used to form the LbL assembly on the PS particles dispersed in water (5% w/v, microParticles GmbH). Aqueous solutions of PEI and PDDA (1 mg/mL) and an aqueous dispersion of mNPs (20 mg/mL) were prepared as the adsorption solutions. 137 μL of PS dispersion was diluted by pure water. 1) The dispersion was centrifuged, the supernatant was removed, and the sediment was dispersed by sonication. 2) To build a layer, 1.5 mL of the adsorption solution was

* Corresponding author: nkato@meiji.ac.jp

 DOI: 10.1051/matecconf/20179805001

added and the dispersion was incubated for 20 min on the vortex mixer to let the material adsorb on the particles. 3) After the incubation, the dispersion medium was exchanged to pure water to remove the excess material by using a centrifuge. The procedures from 1) to 3) were repeated by exchanging the adsorption solutions in accordance with the order of the layer structure (PS particle/PEI/mNPs/PDDA). Finally, the coated particles were dispersed in 1 mL of pure water.

After stirring the mixture of the dispersion of the coated particles (1 mL), cetyltrimethylammonium bromide (50 μmol), H_2O (11 mL), ethanol (4 mL), and aqueous NH_3 (25% w/w, 0.1 mL) for 10 min at 30 °C under N_2 atmosphere, tetraethoxysilane (50 μL) was added to initiate the sol-gel reaction and form the silica/surfactant complex on the coated particles, and the stirring was carried out for another 2 h at 30 °C. The dispersion medium was then exchanged to ethanol by using a centrifuge. The sample was dried and calcined at 550 °C for 6 h in ambient atmosphere to remove the templates.

Pure water (> 18 MΩ cm) used for the mNP synthesis, the LbL coating and the sol-gel reaction was prepared in a Milli-Q system and degassed prior to use.

The ζ-potential and the diameter of the particles were evaluated by laser Doppler electrophoresis and dynamical light scattering (DLS) methods, respectively (Otsuka Electronics, ELSZ). The Hückel equation was used to obtain the values of ζ-potential [23], and the Marquardt method [24] was applied to obtain the distribution of scattered light intensity versus particle diameter. The morphology of the samples was observed using the transmission electron microscope (TEM, JEOL, JEM-1200EXII) by placing the dispersion on the Cu grid covered by collodion film (Nisshin EM). X-ray diffraction (XRD) patterns were recorded on the θ-2θ powder diffractometer (Rigaku, ULTIMA III) using Cu Kα1 radiation. The crystallite sizes were calculated by applying the Scherrer equation to the diffraction peak [25]. The magnetization curves of the samples were recorded on a vibrating sample magnetometer (TOEI, TEM-WF86R-154). The saturation magnetization (Ms) and coercivity field values (Hc) were obtained from the magnetization curves.

3 Results and discussion

3.1. Fabrication of mMSHCs

The ζ-potential was observed after each coating as shown in Figure 1. When the first polycation (PEI) layer was formed on the negatively charged bare PS particles, the sign of the potential was reversed. Because the ζ-potential of the mNPs was −22.6 mV, the sign reversal from positive to negative was observed upon formation of the second mNP layer. Such sign reversal was also observed after coating the third polycation (PDDA) layer. This result suggests that the charge overcompensation occurred upon the adsorption of each layer.

The morphologies of the particles were observed by the TEM. Figure 2(a) shows the bare PS particles and their average dimeter was 202.5 nm. As shown in Figure 2(b), the small black dots densely adsorbed on the smooth surface of the bare particles were observed. Because the black dots were observed after the formation of the PEI/mNP/PDDA layer on the particles, these dots corresponded to the mNPs and the average size of the dots in Figure 2(b) was 10.4 nm. These results indicate that the LbL assembly on the PS particles was successfully achieved.

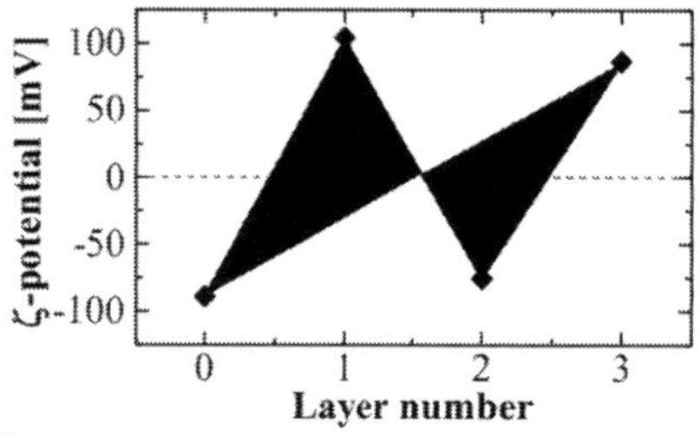

Fig. 1. ζ-potential as a function of layer number. The second layer was composed by the mNPs.

On the particles with the LbL coating, the silica/surfactant complex was formed and the sample was calcined to remove the templates. The TEM image after the calcination is shown in Figure 2(c). The image clearly shows the complete removal of the PS core and the formation of the hollow structure. On the inner surface of the cavity, many black dots, which were similar to the ones observed on the coated particles in Figure 2(b), were observed. Therefore, the mNPs were remained even after the calcination and embedded on the inner surface of the capsules, and the formation of mMSHCs was suggested.

The comparison between mMSHCs and the conventional MSHCs was performed. The MSHCs were prepared under the same condition for the mMSHC synthesis except the use of the PS particles without any coatings. The TEM image of the MSHCs shown in Figure 2(d) exhibits the same morphology as in the previous reports [3-5]. The mMSHCs also show the same shell thickness and the smooth outer surface as shown in Figure 2(c). However, the inner surface of the mMSHCs is not as smooth as that of the MSHCs. This is because the former capsules embedded the mNPs on the inner surface of the cavity. The outer and inner diameters of the mMSHCs and the MSHCs, which were determined by the TEM images, are listed in Table 1. The both diameters of the mMSHCs are larger than those of the MSHCs, indicating that the diameter of the core template for the mMSHCs was larger due to the LbL coating.

The particle size distributions of the bare PS particles and the mMSHCs are also compared in Figures 3(a) and (b). From the distribution, the coefficient of variation (CV) of diameter was calculated, and the values of CV were 26.3% and 43.4% in the bare particles and mMSHCs, respectively. The increase in the CV was comparable to the MSHCs [3, 4]. The structure of the mesopores in the shell was also examined by the XRD. As shown in Figure 4, the mMSHCs exhibited a broad peak at 2θ = 2.56°. Although these characteristics indicated that the periodicity of the mesopores was highly disordered, the same XRD patterns were observed in the

MSHCs [3, 4]. Therefore, not only the incorporation of mNPs in the capsules but also the same structural characteristics as in MSHCs was realized in the obtained mMSHCs.

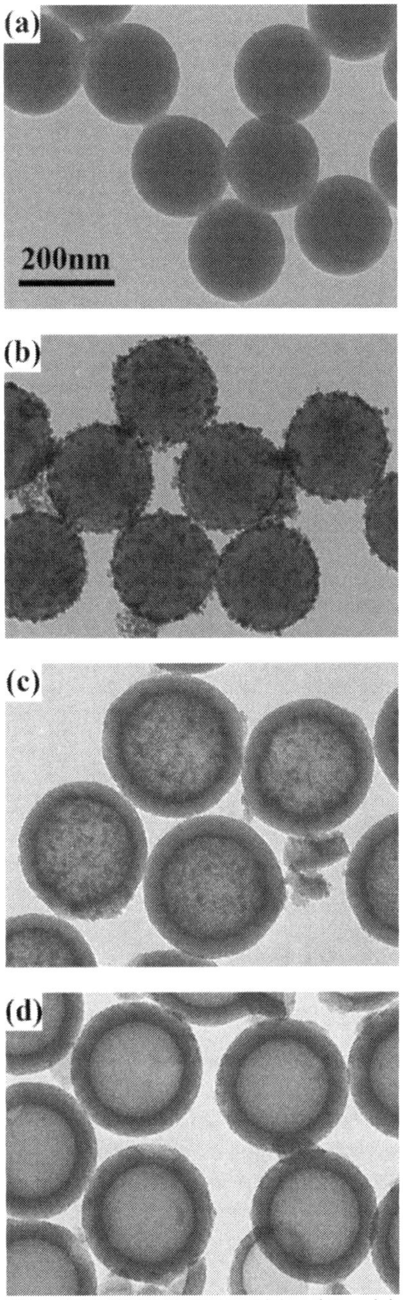

Fig. 2. TEM images of (a) bare PS particles, (b) particles coated by the layer structure of PEI/mNP/PDDA, (c) mMSHCs, and (d) MSHCs. A scale bar for all images is indicated in (a).

Table 1. Sizes of mMSHCs and MSHCs analysed by TEM.

Sample	Outer diameter	Inner diameter
mMSHCs	284.9 nm	210.8 nm
MSHCs	264.4 nm	190.8 nm

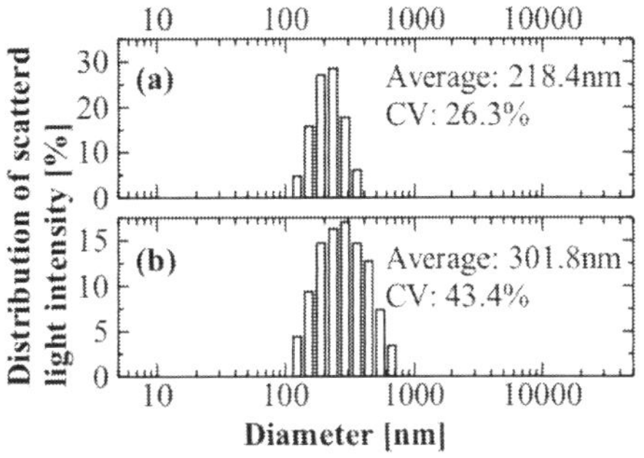

Fig. 3. Particle size distributions of (a) bare PS particles and (b) mMSHCs. Average diameters and CV values were indicated.

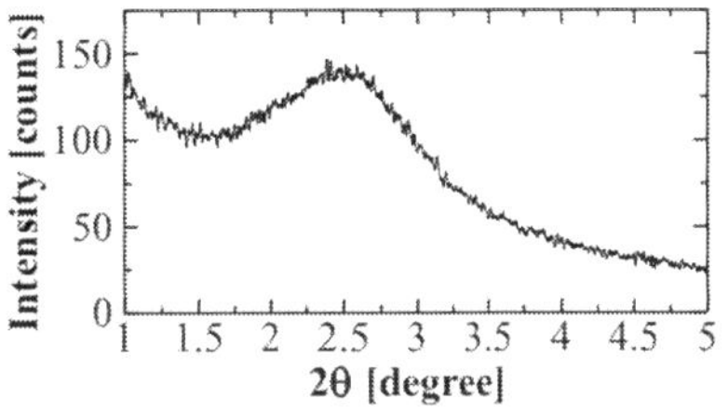

Fig. 4. XRD pattern of the mMSHCs scanned at low 2θ.

3.2 Comparison between mNPs and mMSHCs

The XRD pattern of the as-prepared mNPs shown in Figure 5 indicated that our mNPs consisted of magnetite (Fe_3O_4) crystallites. By applying the Scherrer equation to the (400) diffraction peak, the size of the crystallites was calculated to be 7.8 nm. The mNPs in the mMSHCs also exhibited the same XRD pattern as observed in the mNPs except a broad peak around 2θ = 24°. This peak is the halo pattern of SiO_2 consisting of the shell. By applying the Scherrer equation, the size of the crystallites embedded in the capsules was deduced to be 7.5 nm. Therefore, the magnetite crystallites maintained their size even after the sol-gel reaction and the calcination to form the hollow capsules. The size of the crystallites is comparable to the size of the black dots (ca. 10.4 nm) observed in the TEM images as shown in Figures 2(b) and (c). This similarity in size also supports that the black dots in the TME images correspond to the mNPs.

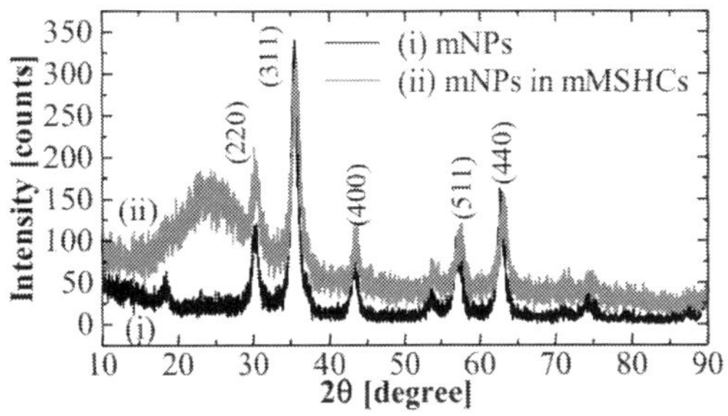

Fig. 5. XRD patterns of (i) mNPs and (ii) mMSHCs.

The magnetization curves of the mNPs and mMSHCs are indicated in Figure 6(a). The Ms values of mNPs and mMSHCs were 57.9 and 14.9 emu/g, respectively. Because the mass ratio of mNPs was lower in the mMSHCs, the reduction of Ms was observed after the formation of mMSHCs. However, our preparation method of the mMSHCs provided the Ms value comparable to the reported magnetic capsules [8-18]. As shown in Figure 6(b), the Ms value of our mMSHCs was large enough to show a magnetic separation using a neodymium magnet. The inset of Figure 6(a) shows that mNPs and mMSHCs exhibited the Hc of 25.4 and 52.1 Oe, respectively, indicating that our magnetite crystallites did not show the superparamagnetic characteristic. However, the Hc value was small enough for the redispersion of the mMSHCs after the magnetic separation, i.e., once the magnet is removed after the collection of our mMSHCs by the magnet as in Figure 6(b), they were easily redispersed by slight shake.

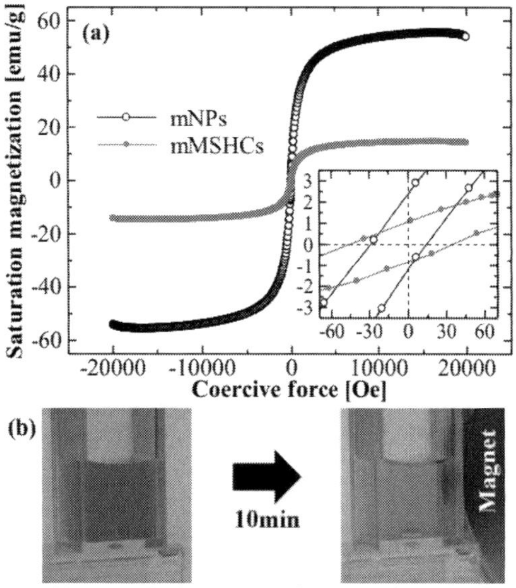

Fig. 6. (a) Magnetization curves of mNPs and mMSHCs. (b) Magnetic separation of mMSHCs using a neodymium magnet.

4 Conclusions

The method for synthesizing the mMSHCs was developed by combining our protocol for the MSHC synthesis with the LbL method to integrate the mNPs onto the PS particles, which act as the templates for the hollow cavity in the MSHC synthesis. The mNPs were successfully incorporated in the inner surface of the MSHC cavity. The structural characteristics of our conventional MSHCs, i.e., the uniformities of the size and shape in the capsules and the structure of the mesopores in their silica shells, were maintained in the resultant mMSHCs. The magnetic separation was also demonstrated. Because it had been demonstrated that the diameter of the MSHC could be reduced less than 100 nm [3], our mMSHCs could be one of the candidates for the magnetically targeted carriers.

Acknowledgements

The authors acknowledge the financial supports from The Institute of Science and Technology, Meiji University and partially from JSPS KAKENHI Grant Number 16K01399.

References

1. Y. Li, J. Shi, Adv. Mater. **26**, 3176 (2014)
2. Y. Bao, C. Shi, T. Wang, X. Li, J. Ma, Micropor. Mesopor. Mater. **227**, 121 (2016)
3. N. Kato, T. Ishii, S. Koumoto, Langmuir **26**, 14334 (2010)
4. N. Kato, N. Kato, Micropor. Mesopor. Mater. **219**, 230 (2016)
5. Y. Yamaguchi, N. Kato, TechConnect World 2014 Proceedings, in: *Nanotechnology 2014*, vol. 2, pp. 315-318 (CRC Press, Florida, 2014)
6. Y. Chen, H. Chen, J. Shi, Adv. Mater. **25**, 3144 (2013)
7. P. Kovačík, M. Singh, F. Štěpánek, Chem. Eng. J. **232**, 591 (2013)
8. J. Liu, Y. Deng, C. Liu, Z. Sun, D. Zhao, J. Colloid Interface Sci. **333**, 329 (2009)
9. F. Wang, Y. Tang, B. Zhang, B. Chen, Y. Wang, J. Colloid Interface Sci. **386**, 129 (2012)
10. J. Zhou, W. Wu, D. Caruntu, M.H. Yu, A. Martin, J.F. Chen, C.J. O'Connor, W.L. Zhou, J. Phys. Chem. C **111**, 17473 (2007)
11. L. Zhang, S. Qiao, Y. Jin, Z. Chen, H. Gu, G.Q. Lu, Adv. Mater. **20**, 805 (2008)
12. F. Liu, J. Wang, Q. Cao, H. Deng, G. Shao, D.Y.B. Deng, W. Zhou, Chem. Commun. **51**, 2357 (2015)
13. W. Zhao, H. Chen, Y. Li, L. Li, M. Lang, J. Shi, Adv. Funct. Mater. **18**, 2780 (2008)
14. Y. Zhu, E. Kockrick, T. Ikoma, N. Hanagata, S. Kaskel, Chem. Mater. **21**, 2547 (2009)
15. Y. Zhu, Y. Fang, S. Kaskel, J. Phys. Chem. C **114**, 16382 (2010)
16. X. Zhang, L. Clime, H. Roberge, F. Normandin, L'H. Yahia, E. Sacher, T. Veres, J. Phys. Chem. C **115**, 1436 (2011)
17. C.X. Lin, Z. Li, S. Brumbley, L. Petrasovits, R. McQualter, C. Yu, G.Q. Lu, J. Mater. Chem. **21**, 7565 (2011)
18. S. Sahu, N. Sinha, S.K. Bhutia, M. Majhi, S. Mohapatra, J. Mater. Chem. B **2**, 3799 (2014)
19. F. Caruso, R.A. Caruso, H. Mohwald, Science **282**, 1111 (1998)
20. F. Caruso, M. Spasova, A. Susha, M. Giersig, R.A. Caruso, Chem. Mater. **13**, 109 (2001)
21. F. Caruso, Adv. Mater. **13**, 11 (2001)
22. Y. Deng, W. Yang, C. Wang, S. Fu, Adv. Mater. **15**, 1729 (2003)
23. I.W. Hamley, *Introductrion to Soft Matter*, Sec. 3.4 (John Wilay & Sons, Ltd., West Sussex 2007)
24. D.W. Marquardt, J. Soc. Ind. Appl. Math. **11**, 431 (1963)
25. A. Guinier, *X-Ray Diffrction*, Chap. 5 (Dover Publications Inc., New York, 1994)

AUTHOR INDEX

EDP Sciences – Web of Conferences
17, Avenue du Hoggar
Parc d'Activité de Courtabouf
BP 112
F-91944 Les Ulis Cedex A
France

ISBN 978-1-5108-3608-2

CURRAN ASSOCIATES INC.
proceedings.com